AEROCAR

AEROCAR

1928–1940

Andrew Woodmansey

4880 Lower Valley Road • Atglen, PA 19310

Cover and title page image: Curtiss posing with an Aerocar built for M. D. Graves of Pittsfield, Massachusetts, with Graves' Cadillac. The photo is taken in front of Curtiss's house in Miami Springs, Florida (1929).

Library of Congress Control Number: 2024931556

Cover design by Chris Bower
Type set in Arboria/Scotch Text

ISBN: 978-0-7643-6809-7
Printed in India

Published by Schiffer Publishing, Ltd.
4880 Lower Valley Road
Atglen, PA 19310
Phone: (610) 593-1777; Fax: (610) 593-2002
Email: info@schifferbooks.com
Web: www.schifferbooks.com

For our complete selection of fine books on this and related subjects, please visit our website at www.schifferbooks.com. You may also write for a free catalog.

Schiffer Publishing's titles are available at special discounts for bulk purchases for sales promotions or premiums. Special editions, including personalized covers, corporate imprints, and excerpts, can be created in large quantities for special needs. For more information, contact the publisher.

CONTENTS

ACKNOWLEDGMENTS

This book would not exist without the vision and persistence of Richard Leisenring Jr., curator emeritus, curator of the Glenn H. Curtiss Museum for over twenty years. Rick offered copious information and images for my first book on the history of recreational vehicles and later asked me if I would like to write a history of the Curtiss Aerocar. I was honored to oblige. Thank you, Rick, for your tireless dedication and service in Curtiss's memory, and best wishes in your retirement.

My thanks are also due to the Curtiss Museum's founders, directors, staff, and volunteers for creating and sustaining an important physical repository of Curtiss machines and memorabilia and to the Curtiss descendants who generously made so many wonderful photographs and other artifacts available for research and public exhibition. Thank you to the many museums and libraries in North America for digitizing important Curtiss records and for supporting this project with additional material. I am grateful to Arthur Jones of The Society of Automotive Historians for identifying many of the automobiles shown in the photos used.

Unless otherwise stated, all images in the book are from the archives of the Glenn H. Curtiss Museum. Special thanks must go to the Miami-Dade Public Library System, Hialeah Public Libraries, the Milwaukee Art Museum, the Louwman Museum in The Hague, the Petersen Museum in Los Angeles, HistoryMiami Museum, and the Library of Congress for supplying additional images.

My love and thanks go to my wife, Christine, for her enduring support in this and my other endeavors in my "retirement."

PREFACE

This book tells the story of the Curtiss Aerocar, a multipurpose trailer invented by Glenn H. Curtiss in about 1927. Nearly one hundred years later, a volume dedicated to this radical vehicle is long overdue. The Glenn H. Curtiss Museum of New York felt that my credentials as a recreational-vehicle historian would make me a good person to write this story, and offered me free run of their archives. As a result of this collaboration, many rare Aerocar photos are published here for the first time. There is new information on the origins of the Aerocar, and new insights into Curtiss's thinking on a wide range of topics from automobile design to welfare capitalism. I hope I have done the man, the machine, and the museum justice.

Aerocars were made in Florida and Michigan. Although the Florida-built Aerocars were generally the only ones entitled to be called "Curtiss Aerocars," this term is used here to cover all Curtiss-designed Aerocars regardless of where they were built. This usage helps avoid confusion with at least two other unrelated groups of vehicles called "Aerocars" in transport history—the automobiles made by the Aerocar Motor Company of Detroit (1906–08) and the Taylor Aerocar, a flying car from 1949.

Other than a brief overview of Curtiss's broader achievements in chapter 1, this is not the place to learn about Curtiss's cycling, motorcycling, engine, aircraft, seaplane, and glider achievements. For these, there are many better places to look.

Since the Aerocar was one of Curtiss's last inventions before his premature death in 1930, it contains the DNA of several previous Curtiss machines. Its multipurpose nature was inspired by the wide range of business and recreational activities that Curtiss undertook during his lifetime. For these reasons, I felt it important to give adequate coverage to four key influences on the Aerocar: Curtiss's early machines, his love (and hate) of automobiles, his first camping vehicles, and his time in Florida. These four topics offer important insights into what the Aerocar was and why it came about. I therefore make no apologies for giving them a chapter each before turning to the Aerocar itself. These early chapters may also shed some new light on underreported aspects of Curtiss's life, especially Curtiss's Florida years (1920–30), which have traditionally been glossed over in aviation-focused Curtiss biographies.

It is easy for a history of this kind to become cluttered with technical jargon, patents, adverts, features, price lists, and corporate information. To help this story flow, most of these components have been separated from the book's main body and placed in appendixes. These include drawings and summaries of the main Aerocar-related patents, sample advertisements, price details, specifications, and galleries of private and business Aerocars. To aid in understanding the technical significance of the Aerocar as an early fifth-wheel trailer, a brief and, I hope, accessible overview of the history of the fifth wheel is included for those interested.

The book is in the main illustrated with wonderful photographs from the archives of the Glenn H. Curtiss Museum. It has been a pleasure and a privilege to bring them out into the open. It is hoped that those who come to this book armed with a good knowledge of Curtiss's achievements in aviation will learn something new about the Aerocar. Likewise, those who already know about the Aerocar may learn something new about Curtiss. The main goal of this book, however, is to create a permanent record of the Aerocar before its story and the few extant Aerocars are lost for good. As a remarkable, multipurpose vehicle that was ahead of its time, the Aerocar deserves a prime place in American transport history.

Glenn Hammond Curtiss (1907)

INTRODUCTION

In a Glenn H. Curtiss obituary published in the *New York Times* on July 24, 1930, the Aerocar does not receive a mention. Such were the man's other accomplishments in naval and general aviation that there was no room to include what for a lesser mortal would have been a life-defining achievement.

The Curtiss Aerocar was a road-going, fifth-wheel trailer built from 1928 until 1940. Its many uses included recreation, commerce, and the transport of passengers, patients, goods, and animals. It was a revolutionary form of transport that set new standards in speed, safety, and passenger comfort. It achieved a number of "firsts": it was the first fifth-wheel trailer to be manufactured in the US for leisure purposes, the first-known fully enclosed travel trailer in the US, the first travel trailer in the US to be built using aircraft principles, and the first-known trailer in the US to use a monocoque design. Its size, ride, and build quality were unsurpassed in the trailer industry of the time. With some justification, it came to be known in the 1930s as "the Rolls-Royce of travel trailers."

The Aerocar played a supporting role in the growth of a number of American industries and cities. During the 1930s, hundreds of salesmen and women used Aerocars to introduce products and services to remote communities across America. By providing ground transportation between airfields, railway stations, and hotels that were linked to new airline passenger services from 1929, the Aerocar played a part in the birth of the US commercial aviation industry. By linking hotels, resorts, and sports clubs in Miami during the 1930s, the Aerocar helped the city's tourism industry to flourish.

Sometimes described as "a millionaire's house on wheels," the Aerocar's origins were humble. Aerocar prototypes helped Florida farmers deliver their fruit and vegetables to customers in downtown Miami in the late 1920s. There were Aerocar ambulances, dental clinics, and even a purpose-built model for a patient with an iron lung. Above all, however, the Aerocar was a luxury road vehicle. It was the flagship of the US travel trailer industry in the 1930s. In part due to the role played by the Aerocar, the US travel trailer industry came to be the largest in the world by the mid-1930s, a position the industry still holds today by some margin.

The Aerocar was renowned for its revolutionary pneumatic hitch and its air-cushioned suspension, but its story is a bumpy one. It was conceived during the Roaring Twenties, when America assumed economic and social leadership from war-torn Europe, and anything seemed possible. It straddled an economic slump after World War I, a real estate boom and bust in Florida, the Wall Street crash of 1929, the Great Depression of the 1930s, and the American travel-trailer-manufacturing crash prior to World War II. It is surprising that it survived even for twelve tumultuous years.

In the following pages, we will learn how the Aerocar became a new form of road transportation. We will learn of Curtiss's contempt for automobile design of the period and why road vehicles should, in his view, be built more like planes. We will learn about "auto-intoxication" and how Curtiss saw the Aerocar as its cure. And we will learn that Curtiss envisioned the Aerocar as the first of a new family of vehicles that would make long-distance road travel fast, safe, and comfortable.

As the last machine to be invented by Curtiss that went into production, it incorporates a lifetime of Curtiss's engineering knowledge. But it contained not just technical wizardry. The Aerocar embodied much of Curtiss's own thinking about transportation and its role in society. As such, it is a fitting symbol of Curtiss's life and a lasting tribute to the vision and skills of a rare man.

An early Curtiss Aerocar with a 1928 Pontiac in front of the Opa-locka administration building, Florida (ca. 1930)

CHAPTER 1

THE MACHINES OF GLENN CURTISS

Fundamentally [Curtiss] possesses all of the requisites for a successful aviator. He is active, [is] cool-headed, has an abundance of nerve without foolhardiness, and thoroughly understands the machine he is operating.

Motorcycle Illustrated, August 1, 1909

Opposite: The Graf Zeppelin at the Opa-locka Naval Air Station near Miami on October 23, 1933, with two Aerocars from the Year-Round Club fleet owned by Henry L. Doherty. *Courtesy of the Gleason Waite Romer Photographs Collection, from the Miami-Dade Public Library System*

The Aerocar was in many ways the sum of all Curtiss machines that preceded it. Some flew, but many didn't. As well as providing a brief introduction to Curtiss, this chapter is about some of Curtiss's early inventions that lent some of their DNA to the Aerocar.

WHO WAS GLENN CURTISS?

Glenn Hammond Curtiss (1878–1930) is the unsung hero of early American aviation. In the early part of the twentieth century, he designed and built many of the country's first aircraft along with the engines that powered them. Many of America's early powered dirigibles used Curtiss engines. He developed North America's most important training plane of World War I (the JN-4D "Jenny") and the world's first planes that could take off from and land on water. He has many aviation firsts to his name, including pilot of the first officially recognized, preannounced, and publicly observed flight in America (1908); the first aircraft to make a takeoff (1910) and landing (1911) from the deck of a ship; and the first flight across the Atlantic (1919). He held the Aero Club of America pilot's license no. 1.

But Curtiss was far more than an aviator. He was just as much at home in land machines as in airborne ones. He built and raced first bicycles, then motorcycles, becoming "the Fastest Man on Earth" in 1907, when he rode a V-8 motorcycle along Ormond Beach in Florida at 134.34 mph (219.4 km/h). He sold, built, or modified automobiles and dreamed up a flying automobile in 1917. He built propeller-driven vehicles for use on land and water, and camping vehicles for his family. In later life, he became a community developer in Florida and diversified into tourism, experimental agriculture, and even moviemaking. The last Curtiss vehicle that went into production before his premature death in 1930 was the Curtiss Aerocar. It was a radical new form of road transportation.

THE CURTISS PHILOSOPHY: FAST BUT SAFE

> Do you notice that there are no birds flying out today?
>
> —Glenn Curtiss, Indianapolis, December 10, 1909

This comment by Glenn Curtiss was made to a newspaper reporter in Indianapolis who had asked him whether that day was a good day to fly a plane. The reply says much about the man. Nothing about wind speed or direction. Nothing about weather forecasts or the ability of Curtiss's planes to handle the conditions. Just the absence of birds. If they were not flying, neither would he.

In his flying days, Curtiss would follow this cautious policy to the letter, sometimes to the great disappointment of the crowds who had assembled to watch the incredible spectacle of a man in flight. For all the adjectives that have been applied to Curtiss over the years, such as "daredevil," "hell rider," and "birdman," he was also careful, for the simple reason that his life depended on it. Curtiss knew from personal experience that air currents were unpredictable, and he would rarely fly if the skies were not calm. As a pilot, Curtiss suffered a number of plane-related injuries, including in August 1909, when his aeroplane was hit by a violent side gust in Rheims, France. Even though Curtiss was thrilled by speed, he was also respectful of it. Lessons learned in the air would be applied later on land. Sometimes called a land yacht by its impatient salesmen, the Aerocar was more accurately an aircraft-inspired road vehicle. As such, the Aerocar was a true hybrid.

As well as incorporating practical features from his flying machines such as lightweight construction, streamlining, and careful weight distribution, Curtiss imbued the Aerocar with comfort and safety. These two features went hand in hand, since an uncomfortable pilot would soon become tired, and a tired pilot would soon become unsafe. Both were desperately lacking in the early planes flown by Curtiss.

THE WIND WAGON (1906)

According to Glenn H. Curtiss Museum curator emeritus Rick Leisenring Jr., Curtiss's transition from motorcycles to aircraft began in late 1903, when he developed a V-4 motorcycle engine for early balloonist Thomas Benbow for dirigible use. In 1904, unbeknown to Curtiss, another balloonist, Thomas Baldwin, purchased a Curtiss V-Twin engine from a Californian agent for use in a dirigible at the St. Louis Exposition of that year. When Baldwin's San Francisco facility was destroyed by the San Francisco earthquake of 1906, Curtiss invited Baldwin to his hometown of Hammondsport to make his dirigibles there.

Curtiss's first flight in a dirigible at Kingsley Flats in Hammondsport on June 28, 1907

The Curtiss Wind Wagon (1906)

The Curtiss Ice Boat (1906)

In 1906, Curtiss constructed a test vehicle called the Wind Wagon. It would help him find out how his motorcycle engine could function better when combined with a propeller. Curtiss put three bicycle wheels onto a wooden frame and attached a long rod to the front-wheel fork for steering. At the back of the wagon, he mounted a two-cylinder V-type gasoline engine connected via a geared fan belt to a 6 ft. diameter propeller. It looked like every youngster's dream of a self-built cart on wheels, except that this one was motorized and left a dust storm in its wake.

> [The Wind Wagon] was a great horse scarer and blows up a great cloud of dust when passing along the road, and will even pull the leaves from the trees where the branches are low.
>
> —Curtiss quoted in *Popular Mechanics*, November 1906

The contraption traveled at 30 miles per hour and was not popular on the streets of Hammondsport due to the commotion it made. Curtiss also developed an ice-going version of the Wind Wagon called the Ice Boat, in part to continue his experiments away from the busy streets of the town—but also just for fun.

These experiments proved their worth, since Baldwin went on to use the Curtiss engine and propeller combination in thirteen of his airships. He agreed to move his airship factory to Hammondsport to be close to Curtiss's own factory. The town has never been the same since.

These strange early contraptions are relevant to the Aerocar story, since they demonstrate how Curtiss began to learn in a practical way about power-to-weight ratios, inertia, friction, wind resistance, and the advantages and disadvantages of three wheels (or skis) rather than four in road vehicles.

They also help to explain why, over twenty years later, the Aerocar was not powered by a propeller.

THE CURTISS JN-4D "JENNY" (1916)

The Curtiss JN-4D, more affectionately known as the "Jenny," is the best-known US aircraft from World War I. It was used primarily as a training aircraft but was also adapted for other uses, including goods and ambulance transport. More than six thousand Jennies were built by Curtiss and seven other manufacturers for the US government, plus hundreds more for Canada and Britain.

We could use any one of a dozen of Curtiss's early aircraft to illustrate the strong design connections between the Aerocar and Curtiss's planes, but the Jenny's construction is perhaps closest to that of the Aerocar. To reduce weight but retain the strength of the Jenny, wooden struts and about 1 mile of piano wire were used in its construction, with an estimated 262 turnbuckles to adjust wire tension. The fuselage was built from a lightweight spruce timber frame but was extremely rigid.

A similar construction method was used on the Aerocar over a decade later. Photos of a partially completed Aerocar show its timber frame and wire construction, over which would be applied an insulating board made from sugar cane fibers called Celotex. This created a monocoque chassis that was self-supporting without the need for a heavy steel base. Turnbuckles could be adjusted over the life of the Aerocar to maintain tautness.

Although by the late 1920s the "strut and wire" construction method had been largely superseded in the aircraft industry, it was retained in the 1928 Aerocar because Curtiss and his builders knew it intimately. It could withstand the shocks transferred from the road with only a small weight penalty. It was also easy and cheap to build.

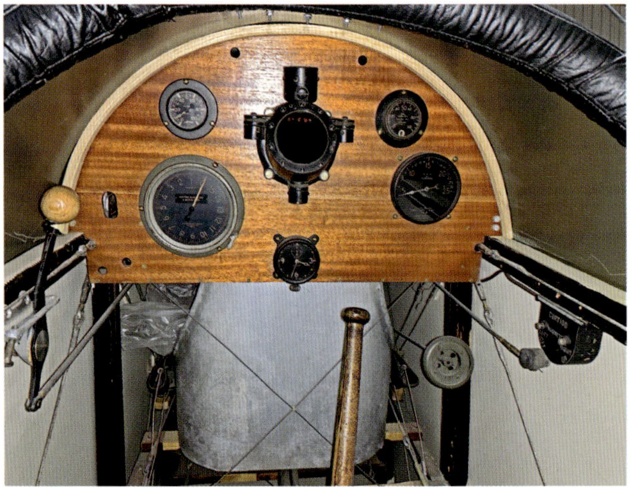

Aircraft features incorporated into the design of the Aerocar

Top left: Wing bracing on a restored Curtiss Jenny JN-4D plane at the Glenn H. Curtiss Museum

Top right: Wall bracing on a Curtiss Aerocar

Bottom left: Cockpit of the restored Curtiss Jenny plane

Bottom right: Observation deck of the Graham Blue Streak Aerocar. *Courtesy of the Louwman Museum, The Hague, Netherlands*

A further nod to aviation construction in the design of the Aerocar was in the observation deck of some luxury models. These were produced after Curtiss's death by two Curtiss engineers, the father-and-son team of Hugh and Harold Robinson. The Aerocar instrument panel mounted into the observation deck was a simplified version of an aircraft cockpit and included a thermometer, altimeter, compass, and clock. Between the instruments were plugs for an intercom system and headset to facilitate communication between passengers and driver. The upper deck of these luxury Aerocars was unique in road vehicles at the time and gave passengers the feeling of being in a plane.

THE CURTISS AUTOPLANE (1917)

Accompanying [the growth in seaplanes] there should be a vigorous development in land machines, chiefly in aerial taxis—a development which will see the aerial runabout as cheap for purchase as the medium-powered automobile—and considerably more economical to run. . . . Think of the ease of upkeep which the aerial taxi will represent. No tire, no bearings, no transmission, no clutch.

—Glenn H. Curtiss, *US Air Service*, August 1919

The Curtiss Autoplane on exhibition in New York (1917)

An artist's impression of the Curtiss Autoplane (1917)

In February 1917, at the Pan-American Aeronautic Exhibition in New York, Curtiss exhibited a four-wheeled triplane that could theoretically use existing road infrastructure to take off and land. Called the Curtiss Autoplane, it was a design exercise dubbed a "limousine of the air" that combined the comfort of an automobile with the freedom of a plane. It was an early experiment in what Curtiss would later call the "aerial taxi," which he envisioned would fly business commuters relatively short distances of up to an hour.

It was also an experiment in comfort. The cabin was fully enclosed and included upholstered seats and window tapestries in an exaggerated attempt to show how comfortably future airline passengers might be accommodated. It was a far cry from the cold, windswept plane cockpits Curtiss was used to. Although its appearance was ungainly, it included some early aerodynamic features typical of later Curtiss road vehicles, such as a domed roof and wheel disks. As far as we know, it never flew, but as an idea for a new form of transport, it was radical. It was the star of the show.

The Autoplane is another Curtiss hybrid. Just like the Wind Wagon, there was no reason in Curtiss's mind why any form of propulsion could not be combined with any form of transport on land, on the sea, or in the air. It is an example of Curtiss's out-of-the-box thinking and his constant search for new ways to travel in speed and comfort.

THE AMBULANCE BOAT (DATE UNKNOWN)

The use of a small boat for the transportation of injured passengers may well have been conceived by Curtiss during the period of high-risk flying experiments of the 1910s as he sought to develop a plane that could take off and land on water. The boat would have been used to rescue the pilot in the event of a crash over water. Other than its eight-cylinder engine, the boat is unremarkable, but this potentially lifesaving marine application foreshadows one of the uses of the Aerocar in the late 1920s as a road-going ambulance trailer. Curtiss's emphasis on safety went beyond vehicle design into thinking about patient recovery—should the worst happen.

HAROLD A. TAYLOR
CORONADO
487

The ambulance boat in use (date unknown)

THE CURTISS SCOOTER (1920)

Curtiss was a keen hunter and fisherman.

When he was trying to reach his favorite hunting and fishing grounds in Florida by boat, he would lose time and patience trying to overcome shallow water and sandbars. To fix the problem, he attached a 400-horsepower aero engine and propeller to the back of a shallow-draft hull. Called the Scooter and christened *Miss Miami*, it was a form of air boat that could travel at 50 mph (80 km/h) over 3 inches of water. It included a cabin with seating for ten people.

Curtiss was not the inventor of the air boat. That honor goes to Alexander Graham Bell with his *Ugly Duckling* of 1905, which was used by Bell at his home in Canada in connection with the flying experiments of the AEA (the Aerial Experiment Association, of which Curtiss was a key member). The Curtiss Scooter is a typical Curtiss refinement, creating something much faster and more practical. Challenging terrain created for Curtiss an opportunity for a new form of transport on water. The leisure objectives of hunting and fishing were the drivers, but the growing importance of seaplanes in naval and commercial aviation offered further scope for invention of a craft that could potentially shuttle passengers between seaplane and shore. For that reason, a Curtiss machine was built to carry a small group of passengers for the first time.

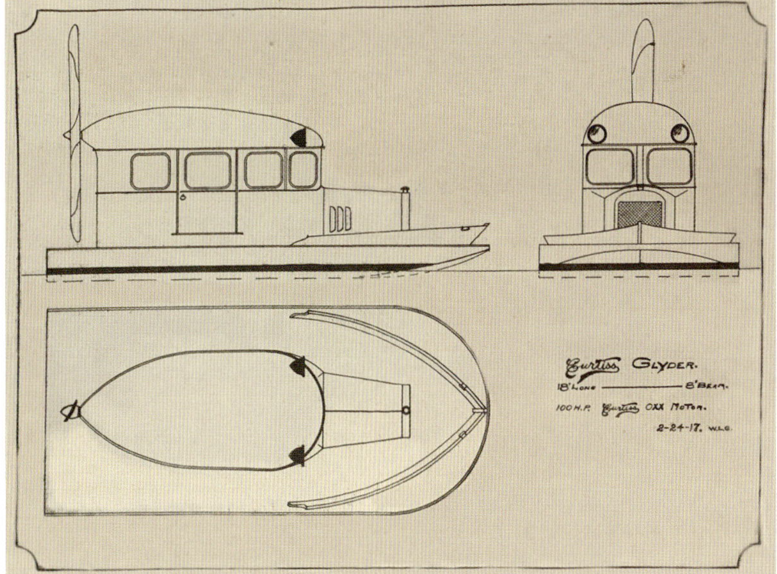

Above: The Curtiss Scooter (1920)

Left: A sketch of a boat called the "Curtiss Glyder," dated February 24, 1917. This is likely to have been an early design for the Curtiss Scooter.

Aerodynamic features of Curtiss planes, boats, and road vehicles

The Curtiss-Wanamaker Flying Boat "America" replica at the Glenn H. Curtiss Museum (1914)

The Curtiss Autoplane (1917)

AN AERODYNAMIC FAMILY

The planes and vehicles developed by Curtiss during the 1910s share a number of aerodynamic features. The curving front of the flying boat "America" (built in 1914 for an attempt to cross the Atlantic, which was abandoned because of the outbreak of war) has been compared to a whale, while the rear of the Autoplane resembles a fishtail. When the "bird-beak" hitch of the Aerocar was added in the late 1920s, we begin to see where Curtiss got his design inspiration from.

As well as designing propeller-driven vehicles for air, sea, and road, Curtiss also explored more conventional automobile design. His experiments with automobiles would likewise contribute to the design of the Aerocar, and it is these vehicles that we shall explore in the next chapter.

Curtiss with his "Scooter" called *Miss Miami* (ca. 1920)

The streamlined rear of an Aerocar (ca. 1929)

"AUTO-INTOXICATION"

One is not comfortable in the motor car. At the end of a 350-mile ride we are fatigued and suffer from "auto-intoxication." The poisons generated by an inactive body make us tired. We cannot move about. We must sit in one position, just as we did in the horse-drawn vehicle 50 years ago[,] and our comfort is not much greater despite the increased speed.

—Glenn H. Curtiss, *Miami Herald*, March 10, 1929

Opposite: Traffic congestion on the County Causeway between Miami and Miami Beach in 1932. *Courtesy of the Gleason Waite Romer Photographs Collection, from the Miami-Dade Public Library System*

A move away from aircraft production into research gave Curtiss time and money to experiment with automobile design. These experiments were important in determining the later appearance and features of the Aerocar.

RICHES AND RESEARCH

In 1916, a restructuring of the Curtiss Aeroplane Company took place in order to address the company's significant wartime order backlog. This presented an opportunity for Curtiss to move away from the manufacturing side of the business, where he was least comfortable. Once a plane was in production, its design was locked in, tying the hands of its creator. Curtiss's main interests were in research and development, and it is here that he was given the chance to focus. Clement M. Keys, a Canadian investment banker who would become Curtiss's financial advisor and friend, saw an opportunity to invest in the rapidly growing aircraft manufacturing sector. He became involved in the company at Curtiss's invitation, when it became clear that Curtiss's small factory in Hammondsport would not be able to keep up with wartime government orders. Keys brought in automobile manufacturer Willys Overland to build aircraft engines and expanded manufacturing to Buffalo, New York. As part of the 1916 restructuring, Keys became president and Curtiss agreed to move to the Curtiss Engineering Company, effectively a research-and-development laboratory where Curtiss could innovate and design to his heart's content, well away from the factory floor. Curtiss reportedly received $4 million as a result of the 1916 restructure. This gave him the opportunity not only to help family and friends financially but also to indulge further in one of his pet hobbies—buying, driving, and, above all, modifying automobiles.

When Curtiss was not in the air or on water, he was on the road. As Curtiss's reputation for flying innovations grew during the 1910s, much of Curtiss's land travel became more mundane, with attendance at flying shows and supervision of his aircraft-manufacturing business requiring significant travel by road or rail. These journeys would have been long, uncomfortable, and tedious.

Curtiss on the roof of his modified Marmon 34. Since Curtiss modified the vehicle to have removable sides and pillars for use in good weather, he may have been conducting informal roof strength tests (ca. 1916).

CURTISS AND HENRY FORD

Curtiss with Henry Ford at Hammondsport with a Curtiss Model F flying boat (1913)

In Hammondsport in 1913, Henry Ford visited Curtiss. Accounts vary as to the reason for this meeting. One Curtiss biographer suggests that Curtiss tried to persuade Ford to put a small but powerful aircraft engine into an automobile, an idea that was apparently politely rejected by Ford on cost grounds. Another account suggests that Ford was building an experimental plane of his own, using a Ford Model T engine, and Curtiss was keen to offer one of his own engines for that purpose. Yet another suggestion is that Ford was interested in buying a Curtiss flying boat. What we do know is that the pair discussed the ongoing patent dispute between Curtiss and the Wright brothers. Ford sympathized with Curtiss's position, having gone through a protracted patent dispute with automobile maker George Selden. Ford offered Curtiss the use of his best patent lawyer. In doing so, Ford is reported to have said, "My entire legal staff is at your disposal. Patents should be used to protect the inventor, not to hold back progress."

It was clear that there was mutual respect between the two men. They came, however, from very different industries, with Ford dedicated to building large volumes of automobiles at low prices and Curtiss experimenting with specialist flying machines that did not belong on a production line. In Ford's case, as it turned out, the allure of aircraft manufacturing would lead Ford into playing a leading role in this industry during and after the First World War. For Curtiss, although the economics and discipline of large-scale automobile manufacturing would not have sat well with his desire to experiment continuously, he would later seek new and better ways of designing automobiles.

CURTISS'S PRIVATE AUTOMOBILES

Mr. D. L. Masson, Curtiss's first automobile customer, with an Orient Motor Buckboard from 1903. In 1907, Curtiss built his own version of the Buckboard, using a two-cylinder motorcycle engine, which never went into production.

Curtiss's interest in automobiles was lifelong, overshadowed by his other transport passions but ever present. In the early 1900s, not content with building bicycles and motorcycles, Curtiss became the Hammondsport distributor for several automobile makers, including Orient Automobile, Frayer-Miller, and Ford. This early exposure to several makes of automobile helped Curtiss understand how automobiles were built and how they might be improved. His earliest-known attempt at automobile construction is a Curtiss Buckboard prototype from 1907.

Curtiss personally owned many automobiles during his life, one report in the *Statesman Journal* of August 17, 1913, putting the number at twenty-seven up to that date. Some vehicles reported by the *Hammondsport Herald* and others as being owned by Curtiss include a Ford (1907), a Peerless Four (1908), a Stoddard Dayton 4 Runabout (1908), an Electric Runabout (1909), a Winton 7 Passenger Touring Car (1911), a Flanders Electric Car (1912), a Chalmers Four (1912), a Jeffery Six (1913), a Buick Touring Car (1914), a Pierce-Arrow Touring Car (1915), and a Smith Motor Wheel (1917). Despite occasional public endorsement of certain automobile brands, Curtiss was not loyal to any single maker, preferring to cycle continually through the latest designs on the market.

THE BRUNN & CO. CUSTOM CURTISS AUTOMOBILES

Between 1912 and 1916, Curtiss commissioned four custom-built automobiles for personal use from Buffalo-based body builder Brunn & Co. In each case, Curtiss made numerous suggestions to the company on how to improve streamlining and comfort. He was closely involved in their design and was a regular visitor to the Brunn factory during their construction. These diversions into automobile design may have offered some light relief from the pressures of wartime aircraft production.

Four automobiles built for Curtiss by Brunn & Co. of Buffalo (1912–16)

Top left: Marmon 34 Chassis with Curtiss modifications to suit all driving conditions. The windscreen was V shaped, sides and pillars were removable, and all windows dropped into pockets for fine-weather driving.

Top right: Pierce-Arrow chassis with heavily slanted rear roof

Bottom left: Cadillac town car. The radiator shell was designed to mimic that of a Rolls-Royce and was engraved with a "C" instead of "RR."

Bottom right: Pierce-Arrow 66 chassis with aerodynamic modifications. The windscreen was again V shaped, the wheels featured wheel disks, and the rear roof was heavily rounded. Curtiss-modified Westinghouse air springs projected front and rear.

THE CURTISS MOTOR CAR COMPANY (1921–22)

After World War I, the problem of what to do with large volumes of surplus aircraft engines prompted Curtiss to explore putting the Curtiss OX-5 aircraft engine into automobiles. According to automobile historian Keith Marvin, the OX-5 engine was used in a small number of automobiles made by Prado Motors Corporation of New York City in 1921–22 before the 1921 economic depression forced the company to close. Dallas, Texas-based Wharton Motors Company was also interested in using the OX-5 but produced only one prototype roadster before they too fell victim to economic circumstances.

In 1921, Curtiss began his own attempt to build OX-5-powered automobiles. The OX-5 engine was modified for road use by Charlie Kirkham, chief engineer of the Curtiss Aeroplane and Motor Corporation. The market was first tested with two conversions—a rebodied 1910 Winton and a Marmon Model 34 cloverleaf roadster. The Winton was owned by C. Roy Keys, a cousin of Curtiss financier C. M. Keys and a vice president of the Curtiss Aeroplane and Motor Corporation.

In a 1921 advertisement in the *Aerial Age Weekly*, the Curtis Aeroplane and Motor Corporation invited aviation fans to travel at speeds up to 100 mph (161 km/h) in either the Winton or the Marmon chassis powered by the OX-5. It seems there were few takers. Despite limited interest in the conversions, Curtiss sought to formalize his initial auto construction efforts by establishing the Curtiss Motor Car Company in 1921. Marvin comments that Curtiss asked Miles Harold Carpenter, designer and manufacturer of the luxurious Phianna motor car, to run the company and build automobiles, with financing to be provided by C. M. Keys. Some initial designs were produced, but this experiment too was cut short by the 1921 depression. The Curtiss Motor Car Company ceased operations in 1922. Surplus OX-5s were sold off as war surplus material for as little as $50. Marvin estimates that no more than fourteen OX-5-powered automobiles were ever made.

Examples of automobiles modified by Curtiss to accommodate the OX-5 aircraft engine (also shown)

Top left: Marmon 34 with OX-5 engine

Top right: C. Roy Keys and his wife in their 1910 Winton with OX-5 engine

Bottom left: The OX-5 installed in the Marmon 34

Bottom right: The OX-5 adapted for automobile use

218 *AERIAL AGE WEEKLY, November 14, 1921*

AVIATION FANS

build yourself a $10,000.00 automobile chassis with an old high-grade chassis and an overhauled OX-5, and have the satisfaction of driving a personally engineered custom automobile.

Very slight and easily made modifications only are necessary ordinarily on the motor and chassis.

The above photograph shows a Winton, year 1910, chassis with OX-5 motor and special body. Weight fully loaded 3,600 lbs., maximum speed 100 m.p.h. Four speed chassis, 18.3 miles per gallon of commercial gasoline by actual continuous check at 30 miles per hour in high gear without slipping clutch, acceleration standing start to 45 miles per hour in 10 seconds. In three days' city driving ran 145.2 miles on 9¼ gallons of commercial gasoline.

Send self addressed and 4-cent stamped large envelope for general data on above and Marmon installation, or one dollar for detail instructions with photographs and OX-5 booklet.

Curtiss Aeroplane & Motor Corporation

GARDEN CITY LONG ISLAND

Advertisement for a Curtiss OX-5-powered Winton Automobile from *Aerial Age Weekly*, November 14, 1921

THE CURTISS AUTOMOTIVE VEHICLE (1925)

In 1925, Curtiss applied for a patent for the Curtiss Automotive Vehicle. It was an invention that showed Curtiss's preoccupation with improving ride comfort by separating an automobile into its propulsion and accommodation components. The intent was to reduce the impact of road and engine noise and vibration on passengers. Comfort was improved by using triangular rather than four-pointed chassis connections. The rear passenger compartment of the vehicle was articulated and sprung above the rear axle of the tractor section. It was a tow vehicle and trailer in a single body—in effect, a triple-axle automobile.

Although the suspension design of this vehicle incorporated a conventional hydraulic shock absorber and rubber discs rather than the pneumatic tire used later in the Aero Coupler, it demonstrated how Curtiss's thinking in vehicle suspension was advancing in the mid-1920s. There are anecdotal reports of an articulated vehicle being seen at a Curtiss airfield in Florida in the late 1920s, but no known evidence of this vehicle remains today beyond the patent.

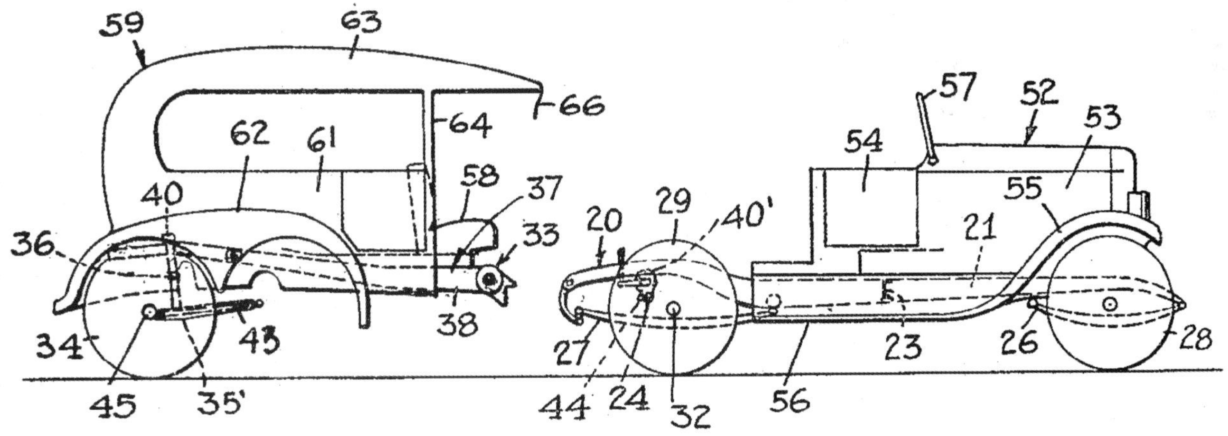

Curtiss Automotive Vehicle patent drawing of 1925

AUTOMOTIVE RESEARCH, INC.

In about 1929, Curtiss formed a small company called Automotive Research, Inc. According to a written company announcement,

> The development work started at Hammondsport last year under the name of "Curtiss Aerocar Company," operating as a branch of the Curtiss Aerocar Co. of Florida, Inc., will be continued as a permanent organization with the same personnel under the name: Automotive Research, Inc. Please govern yourselves accordingly.

Little is known about Automotive Research, Inc., and its tiny presence at Hammondsport. A number of early Aerocars were photographed at Hammondsport in the late 1920s undergoing various design changes. As hinted by the announcement, it's probable that a number of Curtiss engineers who had not followed Curtiss to Florida were helping their southern colleagues refine the Aerocar and design new models and features. It appears then that the name change to Automotive Research, Inc., was a broadening of its initial purpose to reflect Curtiss's desire, not only to refine the Aerocar but to develop altogether new automobiles. This desire appears to have been stimulated by two factors: the success of the Aerocar and the advent of a new automotive technology—front-wheel drive. It is not certain but highly probable that the next two vehicles designed by Curtiss were built by Automotive Research, Inc., at Hammondsport.

Announcement

The development work started at Hammondsport last year under the name of "Curtiss Aerocar Company," operating as a branch of the Curtiss Aerocar Co. of Florida, Inc., will be continued as a permanent organization with the same personnel under the name --

Automotive Research, Inc.

Please govern yourselves accordingly.

Above: An undated announcement changing the name of the Hammondsport branch of the Curtiss Aerocar Company to Automotive Research, Inc.

Right: A small sign (*ringed in red*) on a building in Hammondsport reads "Automotive Research, Inc."

FLIRTING WITH FRONT-WHEEL DRIVE

In the late 1920s, Curtiss became interested in front-wheel-drive automobiles. The January 1930 edition of *Scientific American* summarized the benefits of front-wheel drive as being improved safety, comfort, and economy. These are features that would have appealed to Curtiss in his quest for the perfect automobile design.

The Glenn H. Curtiss Museum archives contain a design, a model, and a photo of automobiles that look remarkably similar. When told to "govern themselves accordingly," employees at Automotive Research, Inc., clearly took this to mean they could do whatever they liked and leave little or no record of their work. There is consequently no written confirmation that these three artifacts are of the same vehicle beyond their visual similarity. We assume that they are the only remaining records of a front-wheel-drive automobile designed by Curtiss and his associates on the basis of a Cord L-29 coupe by Auburn, one of the earliest front-wheel-drive manufacturers. It was reportedly completed shortly before Curtiss's death in 1930. It may have been damaged in an accident shortly thereafter, since it was rebuilt by the Curtiss Aerocar Company in 1933, the date of the only photo of this vehicle. Its steeply raked rear and streamlined front end is reminiscent of the Brunn conversions as well as the Aerocar. Curtiss was reportedly sufficiently impressed with the Cord front-wheel-drive concept to involve the company in the building of the engine for his next vehicle, perhaps the most fascinating Curtiss automobile of all.

THE "AERO HOUSE CAR" (1930)

The dilapidated wooden model of what is presumably the Curtiss/Cord automobile of 1929 was donated to the Glenn H. Curtiss Museum by Henry Kleckler, Curtiss's most trusted, loyal, and highly capable engineer from motorcycling days. Accompanying the model is a handwritten note from Kleckler on his old letterhead that reads as follows:

> The Curtiss Aerocar
>
> This is a model of the Aerocar that I built for G. H. Curtiss in 1930. The Aerocar was a 5-passenger sedan of unit body welded tubing frame construction. It had a front-wheel drive and was powered by a 6-cylinder flat opposed engine designed especially for it. It handled especially well . . . its roadability was excellent.
>
> —Henry Kleckler, April 4, 1992

So here we have a single-bodied, front-wheel-drive vehicle that its builder calls an "Aerocar." Did Kleckler not know that the Aerocar was a fifth-wheel trailer? Of course he did. The Aerocar label for this front-wheel-drive automobile gives us a clue as to Curtiss's automobile design plans, tragically never to be fulfilled. It was in effect the first attempt at an Aerocar "house car." A house car was the term used in the 1920s to describe a motorized truck that incorporated accommodation. In other words, Curtiss was developing a

Top left: Undated, unsigned sketches of an automobile from the Glenn H. Curtiss Museum archives

Top right: A wooden model of an automobile (date of model unknown)

Bottom: Cord L-29 coupe by Auburn, ca. 1929, refurbished ca. 1933

motorized version of the Aerocar trailer. It appears that the Aerocar name was to be continued in a new family of self-propelled vehicles. It was a second front-wheel-drive automobile of much-larger proportions than the first. It used a new, compact engine and was prototyped in 1930 by Curtiss and Kleckler in Hammondsport. It was called in its patent the "Curtiss Motor Vehicle."

> The new power car is ready for the power plant. We will try it out first with a standard power plant and take our time in developing the new motor. Front wheel drives are a great success.
>
> —Letter from Curtiss to C. M. Keys, August 2, 1929

Curtiss's reference to a "power car" in his letter to financier and now friend C. M. Keys is almost certainly refers to the Curtiss Motor Vehicle, for which the patent was applied in July 1929 and granted after Curtiss's death in 1934. The patent makes it clear that to achieve greater passenger comfort, this vehicle is articulated (like the Curtiss Automotive Vehicle of 1925), consisting of a forward section of an engine with a front-wheel-drive steering mechanism and a rear passenger section isolated from road and motor vibrations by using not one but three vertically mounted pneumatic shock absorbers of the Aero Coupler type (the Aero Coupler is described in more detail in chapter 5). A side effect of using a front-wheel-drive engine was the creation of much greater space in the passenger compartment, which could have been put to a variety of uses.

The Curtiss Motor Vehicle prototype (1930)

A Curtiss acquaintance in Florida, Colonel James Prentice, described this radical vehicle in a 1930 article in *U.S. Air Services*:

> During the past two years [Curtiss] was working intermittently on an automobile that had no chassis. It was simply this trailer [the Aerocar] with a front drive and steer axle secured to the framework by three of those pneumatic couplers. He was using one of those modern front drive and steer automobile axles that are so common since the Cord Car came out. As a matter of fact the Cord automobile engineers have done much of the work in the design of this propelling unit.
>
> The engine he was having made when he died is a revolutionary thing. It had horizontal cylinders and it was possible to lift the crankshaft out without even removing the block or engine heads. The fly-wheel was inside the engine, or rather the flywheels were inside, since there were several so that there is no whip. It is a short crankshaft.
>
> —From "The Influence of Aircraft Design on the Trend of Motor Vehicle Construction," by Colonel James Prentice, US Army, retired, in *U.S. Air Services*, December 1930

The Curtiss Motor Vehicle was innovative in its design and engineering. To be clear, the term "Aero House Car" is the author's; this phrase was never used at the time, either by Curtiss or his designers. But from Kleckler's description of its predecessor, there is little doubt as to this vehicle's origins and purpose. It was to be a motorized version of the Aerocar trailer that would be both quiet and spacious. Its streamlined styling was almost a decade ahead of "one-box" leisure and business vehicles designed in the late 1930s, such as the Hunt Housecar of 1937 and the 1938 Western Clipper of industrial designer Brooks Stevens. It is also an early and arguably far more attractive forerunner of the campervans of the 1950s, typified by the Volkswagen Kombi. The recreational-vehicle industry of the US, and indeed the world, may have looked very different in the 1930s if the Curtiss Motor Vehicle of 1930 had gone into production.

AEROCAR FOUNDATIONS

Curtiss's forays into automotive design and manufacturing were attempts to improve a product that he saw as deeply flawed. Curtiss's relationship with the automobile had been initially positive. In contrast to the buffeting he received in his early planes, Curtiss welcomed the temperature and noise insulation offered when cocooned inside an automobile sedan.

> As a pleasure craft the invention seeks to introduce into the aeronautical art the many comforts and conveniences now present in the automobile art.
>
> —Curtiss Autoplane patent application, February 14, 1917

These "comforts and conveniences," compared to early aircraft, included side-by-side seating for better communication with a passenger, greater internal space, and better insulation.

During the 1920s, however, Curtiss's patience with automobile design wore thin.

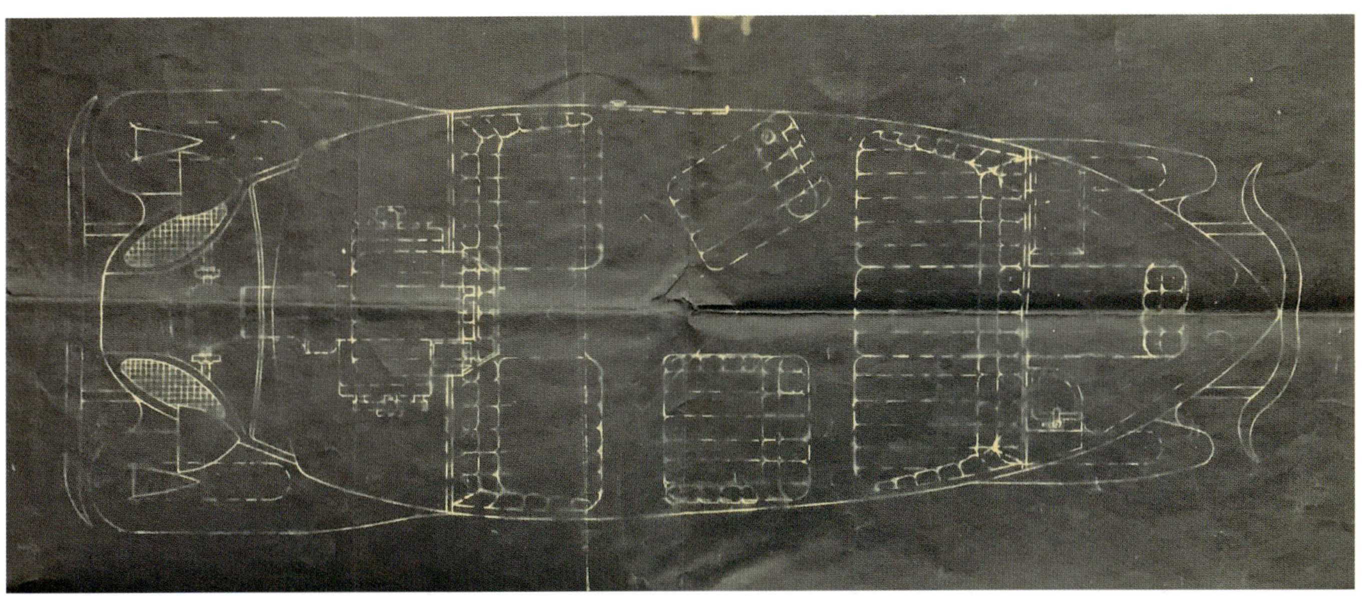

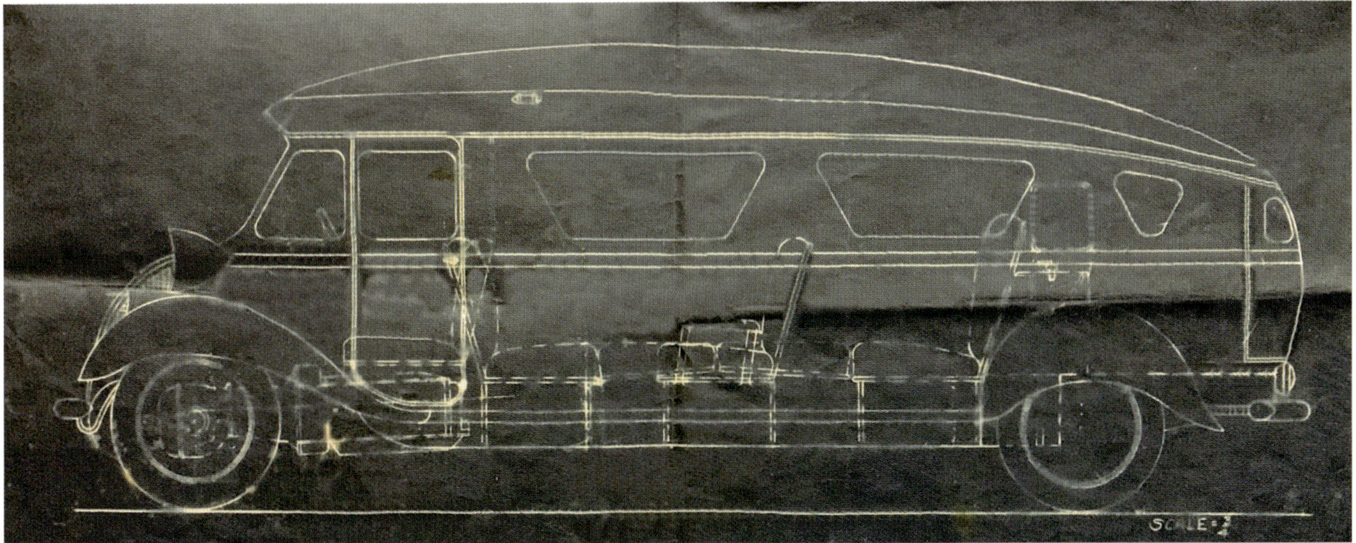

Two blueprints of the Curtiss Motor Vehicle prototype, showing its profile and a proposed seating configuration. It was probably the first streamlined house car ever designed in America (ca. 1930).

"AUTO-INTOXICATION"

> [Curtiss] seemed to have a perfect horror of the engine and road vibrations that are so common while operating automobiles.
>
> —Colonel James Prentice (1930)

Curtiss was not a man of many words. Probably his longest newspaper interview came toward the end of his life, in 1929. Remarkably, it was not about planes or even motorcycles, but automobiles. In the interview, we infer that the Aerocar was in part built in response to what Curtiss saw as the failings of automobile design. Growing frustration with automobiles had turned into thinly disguised contempt, a sentiment Curtiss rarely expressed in public. Here's just part of what he had to say to the *Miami Herald* on the subject in March 1929:

> Man can move on the ground in almost anything. It is the easiest thing to do. Even the cavemen enjoyed log-wheeled transportation. But when it comes to staying in the air[,] man has had to plan for one thing only—efficiency.
>
> Compared with the airplane the automobile is a horrible example of inefficiency. The automobile today is in some respects not far advanced from the horse and buggy stage. All it has gained in 20 years is power and reliability. It is no safer than the horse and buggy of 50 years ago.
>
> We have been grossly extravagant in motor car design. We have done little to economize. With a 56-inch tread, a gargantuan and cumbersome motor, over a ton of steel[,] and a kaleidoscopic array of nickel trimmings, wheels, oversize fenders, lights and fittings, all taking up enough space to house a Nairobi family, it still costs more to travel per mile than it did with a four-horse team and coach a century ago.
>
> One is not comfortable in the motor car. At the end of a 350[-]mile ride we are fatigued and suffer from "auto-intoxication." The poisons generated by an inactive body make us tired. We cannot move about. We must sit in one position, just as we did in the horse-drawn vehicle 50 years ago[,] and our comfort is not much greater despite the increased speed.
>
> I believe there will be a complete revolution in automobile design very shortly. It will be based on a study of present-day aeronautics.
>
> —Glenn Curtiss, *Miami Herald*, March 10, 1929

In these words we begin to see the "why" of the Aerocar. If we view it as everything that Curtiss believed the automobile of the time was not, we are closer to understanding the reason for its creation. We can also see in these words a visionary forecast of the streamlining movement that was to sweep across automobile and recreational vehicle design around the world in the 1930s.

A NEW FORM OF ROAD TRANSPORTATION

If there is one primary reason for the development of the Aerocar, it was as a remedy for "auto-intoxication." Curtiss believed that noisy, stuffy, and cramped automobiles led to passenger fatigue. In traveling many thousands of land miles by road and rail during his flying days between airports, railroad stations, hotels, and home, Curtiss would have been extensively subjected to such discomforts.

By the end of the First World War, planes were ready to travel reasonable distances carrying small numbers of passengers. In Curtiss's view, if road-going vehicles were going to compete with trains and planes over long distances, they would need to be fast, spacious, and comfortable. Curtiss had lost a number of friends and colleagues to plane crashes in the 1910s and knew the risks that came with speed, so safety would also be essential.

The fully fledged Aerocar of 1928 onward was Curtiss's answer to these problems. It was not just a recreational vehicle or a goods trailer or passenger bus. It was a new form of road transportation. Although its conception came about in the midst of the free-wheeling, dealmaking, fun-loving, and mildly chaotic background of 1920s America, its purpose was serious—to make long-distance road travel fast, comfortable, and safe.

If Curtiss had lived longer, we would almost certainly have seen further initiatives in this area. The Curtiss Motor Vehicle of 1930 gives a tantalizing glimpse of how the Aerocar was evolving in Curtiss's mind from a trailer into a self-powered vehicle.

But the immediate forerunner of the Aerocar had a less ambitious goal—to take Glenn Curtiss and his family camping.

CHAPTER 3

THE ADAMS MOTORBUNGALO (1920–22)

In 1920, a second corporate restructuring offered Curtiss the chance to leave aircraft manufacturing altogether. He would take up many new challenges, including designing and building his first camping vehicles. The Adams Motorbungalo was named after Curtiss's half brother but was a Curtiss design. It was the start of a family of Curtiss leisure vehicles that would lead to the Curtiss Aerocar.

Gypsie Life—Modernized!

—From the first-known Adams Motorbungalo advertisement, July 1920

Opposite: The Adams Motorbungalo with a Ford Model T (1922)

The Curtiss Long Island plant, ca. 1919. The small trailer in the center of the photo would later have a role to play in the development of the Adams Motorbungalo.

POSTWAR RESTRUCTURING

By the end of World War I, Curtiss-related companies had manufactured about ten thousand aircraft, eight thousand of which were for the US government. It had been a profitable period both for Curtiss's companies and for Curtiss personally. But there was a cost to this success. In contrast to the corporate restructuring of 1916, which dealt with a backlog of aircraft and engine orders, a second restructure came in 1920 to deal with the problem of having no orders at all. Immediately after the armistice in November 1918, the US government canceled its military supply contracts, including many for Curtiss aircraft and engines that were still under construction. Curtiss Jennies in their hundreds were sold off as surplus stock, becoming in the process the workhorse of the new American form of aerial entertainment known as barnstorming. There was no revenue to pay debts or wages, and no immediate sign of new aircraft orders coming from the private sector. The Curtiss Aeroplane and Motor Company was in trouble.

In September 1920, despite the cancellation of government contracts, financier and Curtiss supporter C. M. Keys felt that aircraft manufacturing had a bright future. He was willing to keep investing in aircraft manufacturing, accepting some short-term pain while the industry recovered. Keys knew that Curtiss was no longer interested in manufacturing but was keen to remain involved in design. So Keys personally bought out the interests of automobile maker Willys Overland, which had been brought in to meet wartime demand, stayed on as president, and kept Curtiss as a technical advisor. Curtiss sold his remaining interests in the company for a reported $32 million, and Keys went on to become a significant participant in the commercial aviation sector as head of the Curtiss Aviation and Motor Company.

This second corporate restructuring was bittersweet for Curtiss. As aircraft and engine production slowed to a trickle, thousands of workers lost their jobs, including employees who lived in Curtiss's beloved hometown of Hammondsport. Many of Curtiss's employees had become friends, and he felt their redundancies deeply. In the last decade of his life, Curtiss's business decisions were in part influenced by a desire to create new jobs, homes, and even lives for some of those whom he felt he had let down.

Since 1916, Curtiss had become less involved in the manufacturing side of the business. A model in production was a model that could not be changed. Later in life, he opened up about this period of his career, saying that the manufacturing focus during the war had been all about brute speed and that the "stick and wire" type of aircraft construction was overtaken by newer construction methods, notably using aluminum. It was the age of the slide rule rather than the blackboard. Planes were now tested in a wind tunnel rather than on a windswept hill or calm lake. Curtiss felt that the aircraft manufacturing industry was "passing him by."

> The war has done little to help progress in airplane building from a commercial standpoint, as there was no demand for safety but for speed and the development of better engines, but the mail service is going to help develop the commercial airplanes.
>
> —Glenn Curtiss, *Tampa Tribune*, April 15, 1920

Churning out planes in their thousands for the war effort had clearly been of limited appeal to Curtiss. There was no experimentation, no improvement, just mass production. He had to look elsewhere to satisfy his need for invention.

This pivotal time in Curtiss's life is important because it is often described as the start of Curtiss's "retirement." This was a dangerous word to use around Curtiss. More accurately, 1920 was the start of a new phase in his life where he would undertake a wide range of projects that brought him personal satisfaction. In the process, he would become busier than ever. He would share his financial success with others and build new communities and businesses in another part of the country. And he would continue to invent new forms of transport.

FROM FLYING TO CAMPING

One of Curtiss's first postwar projects was connected to something close to his heart—camping. Curtiss was used to sleeping rough, since his early-morning starts, when flying conditions were most likely to be benign, often meant overnighting close to an airfield rather than in a comfortable hotel in town. Curtiss was also a keen hunter and needed to stay close to his hunting grounds. His wife, Lena, was keen to explore but less disposed to camping out. According to one report, it was Curtiss's much-loved half brother, G. Carl Adams, who first came up with the idea to make camping more comfortable for the family. In a 1920 article, we learn that

> Glenn H. Curtiss, of airplane fame, and G. Carl Adams are half-brothers; inveterate hunters, fishermen, campers and all-round outdoor men. Roughing it in vacation time is their hobby. Mrs. Curtiss and Mrs. Adams, their wives, likewise feel the urge to roam in season under Nature's own canopy of sky, but for years they were unable to accompany their husbands on extended outings because of the discomforts involved. And how they hated being left behind! They almost wished at times they were men, able to tote gun, rod and tent, and grin and bear everything.
>
> About two years ago Adams said to Curtiss, "Glenn[,] it's a darned shame we have to leave the women-folks at home every time we hark to the call of the wild. Can't we figure out some scheme to make them comfortable, and enable them to enjoy the gypsy life as much as we do?"
>
> —Harry Thompson Mitchell, *Printers' Ink Monthly*, November 1920

So began a journey to design and build a new camping vehicle. According to Mitchell, neither Curtiss nor Adams had set out to turn building a camping vehicle into a business; it was purely to meet personal needs. They spent some time developing possible designs for their camping car before coming up with the first prototype in 1919.

G. CARL ADAMS

G. Carl Adams was Curtiss's half brother. Born in 1897 to Lua Curtiss following her remarriage to J. Charles Adams, Carl Adams was nineteen years younger than Glenn Curtiss. Adams looked up to Curtiss as more of a father figure than a half brother, eagerly following Curtiss's flying exploits as a child. In return, Curtiss would take Adams under his wing and mentor him as long as he lived, in a wide range of Curtiss pursuits from running a business to hunting. At the age of nineteen, with Curtiss's guidance, Adams began

Curtiss (*left*) and Adams (*right*) with similar poses next to a modified Packard automobile (ca. 1920)

producing utility trailers at Hammondsport. In his early twenties, Adams was given the responsibility for manufacturing and selling Curtiss's first production camping car. The new vehicle would be named after Adams. Later, in 1922, Adams would follow Curtiss to Florida and become his right-hand man in a wide range of roles, including passenger bus building, real estate development, construction, and Aerocar manufacturing. After Curtiss's death in 1930, Adams served as mayor of Miami Springs from 1930 to 1942. A memorial to Adams still stands in Miami Springs.

EARLY CURTISS/ ADAMS HOUSE CAR CONCEPTS

It is commonly thought that Curtiss had from the start been interested exclusively in a trailer configuration for his camping car. But a design shared by Adams with readers of *Outers' Recreation* magazine of August 1921 indicates that Curtiss and Adams had at one point considered a truck-based "house car," the name given at the time to a motorized vehicle with integrated accommodation. The article, written by George W. Sutton Jr., showed plans for two camping cars for those wanting to self-build such vehicles. One set of plans came from G. Carl Adams. Sutton commented that both cars "have been built and proved successful." No other records remain of this house car, nor do we know why the house car design was not pursued. It is possible that the vehicle proved too heavy for the sandy or boggy terrain that Curtiss and Adams needed to cross to go hunting. In the mid-1920s, other house car makers would produce motorized vehicles with a similar foldout bed design, such as the Zagelmeyer Kamper Kar of 1925, but this was not the route followed by Curtiss. He went instead with a trailer, but no ordinary one.

WHY A TRAILER?

Why, in 1919, would Curtiss have chosen a trailer for his recreational-accommodation needs rather than an automobile or truck adapted for sleeping? In 1930, Colonel James Prentice, early balloonist and acquaintance of Curtiss in Florida, offered a fascinating insight into this question:

> [Curtiss] had noticed that when they hauled airplanes along the roads by putting the landing skid of the tail end onto a truck, the tools etc. in the airplane fuselage had shaken very little and that men who rode on the planes said there was very little vibration. Some even fell to sleep while being hauled from shops to fields and vice versa.
>
> —From "The Influence of Aircraft Design on the Trend of Motor Vehicle Construction," by Colonel James Prentice, US Army, retired, in *U.S. Air Services*, December 1930

Curtiss had learned that towing a plane backward with its rear end sitting on top of (rather than behind) the truck was a smooth and even sleep-inducing experience. A truck towing a plane had in effect formed a basic fifth-wheel trailer. Prentice went on to explain that the soporific effect of sitting in an aircraft being towed in this way was particularly pronounced after the introduction of balloon tires on aircraft from the mid-1920s. This further observation may have sparked the idea in Curtiss's mind of using aircraft tires instead of automobile tires in his pneumatic coupler.

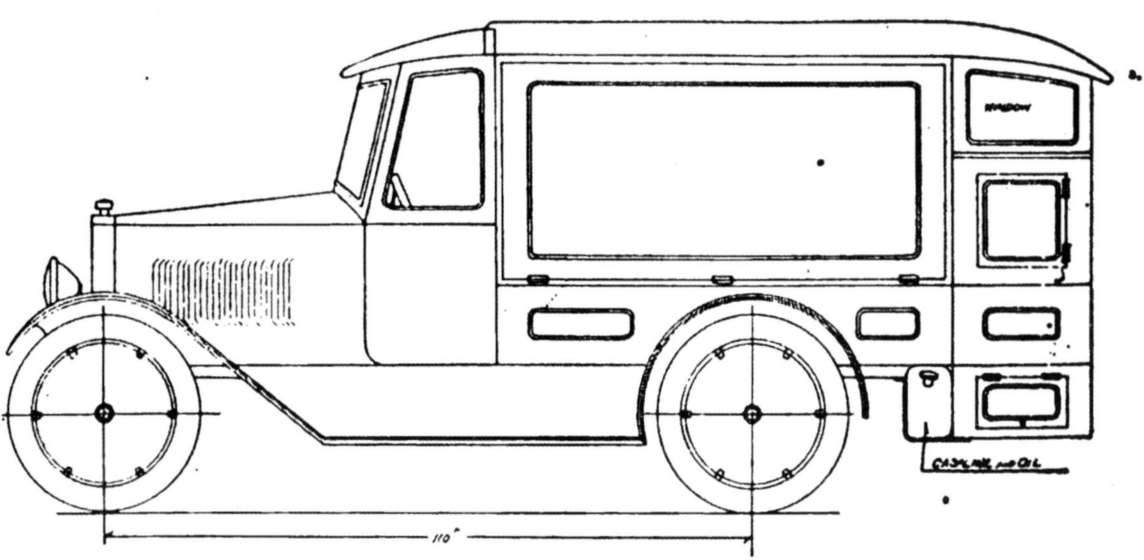

A side view of the interesting Adams car.

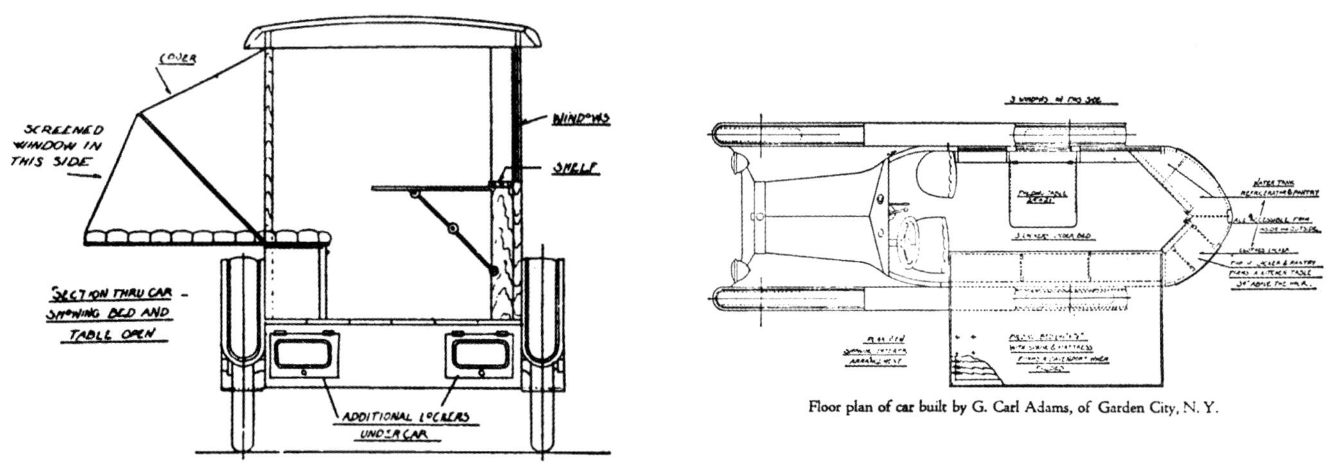

A sectional view of the Adams car.

Floor plan of car built by G. Carl Adams, of Garden City, N. Y.

Three illustrations submitted by G. Carl Adams for a camping car to *Outers' Recreation* in 1921

Towing aircraft and engines between factory, lake, and airfield was a common task in the early days of flying (1912).

Applying lateral thinking to his multidisciplinary transport knowledge, Curtiss was transferring ideas he had seen on the airfield to a new form of comfortable transport on the highway. It was an accidental but significant crossover between aircraft and road vehicle design, identified by a man with great powers of observation.

THE CURTISS CAMP CAR (1919)

The broad design that Curtiss and Adams settled on for their personal recreational vehicle was a trailer with several interesting features, including foldout beds, a tapered front, and a fifth-wheel hitch. This was almost certainly the first fifth wheel used on a recreational vehicle in America. In the patent for Curtiss's Camp Car (see Appendix A), it is stated that Curtiss would submit a separate patent application for the fifth-wheel component, but no such application is known.

We know quite a lot about the prototype's other features, since it featured in the prepublicity of the 1920 production version. The prototype's walls were made of 3⁄16-inch aircraft veneer. Other facilities included an icebox, a water tank, a folding table, a stove and kitchenette, clothes closets, and electric lights. Beds folded out on either side of the car, allowing it to sleep up to six people. A phone was installed to connect the trailer's occupants to the chauffeur, who was provided with a separate sleeping cot. Despite Curtiss's use of aircraft materials, the total weight of the trailer was still 1,800 pounds (816 kg), a significant weight for a trailer of that period.

A close-up of the trailer hitch on the Curtiss Camp Car prototype (*left*) appears to indicate the use of a commercially available Martin Rocking Fifth Wheel (*right*, *from* Vehicle Monthly, *December 1917*).

The first Curtiss Camp Car prototype was rudimentary, as evidenced by the logs supporting the front of this snowbound trailer while a tire was being changed.

The Curtiss Camp Car was used for hunting trips, with photos of the car sometimes showing deer strapped to the side of the tow vehicle (a ca. 1919 Cadillac with shortened body to permit towing). In this photo, G. Carl Adams is on the left. The man on the right is possibly K. B. McDonald (ca. 1919).

"Pullman-type" beds folded out from each side of the camp car. The chauffeur's outdoor cot can be seen on the left (ca. 1919).

The camp car's interior, looking toward the front with beds unfolded. As a prototype, the interior was basic but functional (ca. 1919).

Curtiss and his hunting friends with the camp car in the Adirondacks in October 1919. Always the experimenter, Curtiss (*third from left*) is shown here testing different hunting footwear on each foot. The rear of the photo lists the members of the group as (*from left to right*) Mr. Frank, Carl Adams, G. H. Curtiss, K. B. McDonald, D. C. Merrill, and Ned Ballard.

The concept of fold-down beds in road vehicles was not new. They had been used on the railroads in Pullman Sleeping Cars since they were first patented in 1864, and it is probably this heritage that Curtiss and Adams were paying homage to when later describing their prototype as "of the Pullman type." The prototype had been nicknamed a "motor bungalow" by a family friend. Although this term had been used by others as early as 1911, the name stuck for the prototype. Later, it would be modified for the production models to either "Motor Bungalo" or "Motorbungalo," with Adams's name preceding both.

BUILD ONE FOR ME, PLEASE

Mitchell suggests that the first trip of the prototype was to the Adirondacks "on a sunny day in spring" (presumably in 1919), followed by more trips by "Curtiss and his associates" to the Berkshires and Catskills. By the end of the summer, Curtiss had apparently received many letters from envious campers who had seen the trailer and wanted one too. Mitchell continues by reciting a conversation between Curtiss and Adams:

> "Carl," grinned Curtiss to Adams, "it looks like you are being forced into another business. It was mostly your idea, you know, and, anyhow, I've got to stick to airplanes. Fancy! G. Carl Adams, Builder of Homes for Gypsies De Luxe! Like the idea?"
>
> —Harry Thompson Mitchell, *Printers' Ink Monthly*, November 1920

Adams did, so the pair set to work on a preproduction model. We know that Curtiss was not happy with the first trailer prototype's functionality, but it is not known if a further prototype was built or if the first prototype was modified. Adams explains in a 1922 Motorbungalo brochure:

> At first Mr. Curtiss' camp cars followed in general plan the private Pullman cars familiar to our railroads. They were elegant, luxurious, if you like, but when it came to camping or fishing trips in the sandy or muddy hinterlands of Florida or Georgia, they were impractical because of their size and weight. Attempts to lighten the Pullman type of camp-car resulted in a loss of structural integrity.

Although the design was heavy, Curtiss still decided to patent his prototype camp car. US patent number 1,437,172 was applied for on April 28, 1921, and granted on November 28, 1922 (see Appendix A for the patent's key features). For the production model, Curtiss went back to the drawing board. His next design would be quite different from the prototype.

THE ADAMS MOTORBUNGALO (1920)

The chassis for the revised trailer design came from another enterprise run by Adams. Since about 1917, Adams had been building utility trailers under Curtiss's guidance. These trailers were used to support the production of aircraft at the Long Island plant. The Adams trailer was also sold to farmers and tradesmen. This outwardly unremarkable trailer was a workhorse at the aircraft factory and was especially strengthened to carry heavy aircraft engines. Its V-shaped front end has the hallmark of a Curtiss design, since it was both streamlined and practical, allowing the trailer to be hitched close to the tow vehicle without impacting the tow vehicle's turning circle. The utility trailer's design was novel enough to receive its own patent.

The production version of the Adams Motorbungalo in 1920. This was one of a number of photos taken that year for publicity purposes, featuring actors hired for the purpose.

The Adams utility trailer at work on the farm, date unknown. To reduce weight, Curtiss converted his camp car to a camping trailer. He reduced its length, simplified its interior, and incorporated a mechanism that lifted the roof as the external folding beds were lowered. The reduced-height trailer was lighter and easier to tow, had improved fuel economy, and could be towed by most automobiles. The fifth-wheel hitch of the prototype was replaced with a conventional ball-and-socket hitch.

Three different Motorbungalo models in 1920

Production versions of the Motorbungalo used a conventional ball-and-socket hitch to simplify attachment and widen appeal.

Adams's utility trailer became the chassis for the remodeled camping trailer. In another important design change intended to broaden the appeal of the camper, the entire camping body could be removed from the trailer, leaving owners with a handy utility trailer.

Despite its utility trailer foundations, the production model of the Adams Motorbungalo was in all other respects luxurious. The 1921 and 1922 brochures list four variants based on two main designs—the Model M Closed Motorbungalo De Luxe and the Model R Closed Folding Motorbungalo. A Model S was the same as the Model R with modifications, while a Model T was the same as the Model S but with a kitchenette and wardrobe. A Motorbungalo Junior was added to the range, which was a simpler camping trailer and tent. The number of available models was increased in 1922.

The extensive facilities of the production Motorbungalos included two double spring beds and mattresses, a kitchenette accessible from inside or outside, an icebox, a folding table, a shelf, utensil and dish holders, a water tank, a gas stove, a wardrobe, five windows, a watertight roof made of timber, duck canvas and Fabrikoid, adjustable corner legs, and a rear step. An extensive range of optional camping equipment was made available for sale to accompany the trailer, including a folding toilet and tent.

Importantly, the Adams Motorbungalo weighed only 900 pounds (408 kg). Its conventional ball-and-socket hitch allowed it to be towed by a wide range of vehicles. Curtiss had retained stability and flexibility to travel on soft surfaces by cutting the prototype's weight in half. Adams patented the design of the production model in Canada.

ADAMS MOTORBUNGALO ADVERTISING (1920–22)

Preparations for the public launch of the new camping trailer were made during the first half of 1920. Articles including photos of the first prototype designed by Curtiss were printed in a wide range of publications, including *Automobile Journal*, *Scientific American*, and *Vanity Fair*. The trailer's accommodation was compared somewhat overenthusiastically with the Biltmore Hotel in New York, and significant improvements were promised for the production model. Attaching Curtiss's name to the trailer's design was bound to attract publicity and generate sales.

It was strange, therefore, that the first advertisement for the trailer in July 1920 was rather amateurish. Under the heading of "Gypsie Life—Modernized!" was a sketch of the prototype and the unassuming model name "camp-de-luxe." Interested parties were asked to contact Adams personally.

An Adams Motorbungalo forms part of a romanticized retail camping display (1921).

By October 1920, Adams had become more organized, announcing the name of the trailer as the "Adams Folding Motorbungalo" (the word "folding" was added intermittently throughout the trailer's life) and signing off as the Adams Trailer Corporation of Garden City, Long Island, New York. The Adams Trailer Corporation was incorporated on December 9, 1920, at 125 East 46th Street, Grand Central Palace, New York.

The Adams Motorbungalo was exhibited at the first Highway Transportation Show in New York in January 2021 and featured in a number of automobile showroom camping displays. By mid-1921, Adams's advertising had become more sophisticated, with clear messaging about the benefits of a camping vacation for the busy professional. Examples of Adams Motorbungalo advertisements can be found in Appendix B.

The interior of the Adams Motorbungalo was once again simple but practical (ca. 1921).

MOTORBUNGALO PRODUCTION (1920–22)

Production of the Motorbungalo began in early 1920. In the 1921 Motorbungalo brochure, Adams boasted of "establishing a manufacturing plant . . . with capacity of thirty-five hundred trailers for this year. Part of this output has already been sold and the Corporation is assured of a demand far in excess of its production."

This would turn out to be wishful thinking. The first Motorbungalo models were made at rented factory space in the Lawrence Sperry Aircraft Co. facility in Farmingdale, as well as at Curtiss Engineering Co.'s own factory in Garden City, Long Island. As production increased, the trailers were made in their own plant at Long Island. In the March and April 1921 editions of the *Freeport Daily Review*, Adams advertised for positions at the factory, including woodworkers and sewing-machine operators.

Early sales were promising. Motorbungalos were available for hire as well as for sale, so that potential buyers could try out this new form of camping. The use of the camping trailer by businesspeople and celebrities was publicized in the press. Advertising included appeals for local distributors in a wide range of territories.

Pricing for the Motorbungalo ranged from $485 for a basic Motorbungalo up to $1,200 for a Motorbungalo

Fishing with a Motorbungalo; people, date, and place unknown

Deluxe. The Adams utility trailer cost $195 for the Model A trailer, with several other models available. The utility trailer was sometimes mentioned in the same advertisement as the Motorbungalo and always received a mention in Motorbungalo brochures. This was an interesting decision, given that the target market of the camping trailer was aspiring wealthy campers who may not have been keen to sleep in a utility trailer.

In September 1921, Adams's trailer production was moved from Long Island to Hammondsport to accommodate "demand for trailers and camping outfits from persons planning to winter in the south" (*Motor Age*, September 22, 1921).

Reviews of the Motorbungalo were glowing, such as this one from F. E. Brimmer:

> It is a camp that you may depend upon as dust-proof, bug-proof, and dry. Recently the author and his family, including the two small children, slept in this outfit during a period of rainy weather that lasted over a month and all but took the joy out of Autocamping. However, day and night, we ate, lived, and slept in a bone-dry place inside this outfit.
>
> —From *Autocamping* by F. E. Brimmer (1923)

However, by 1923 it had become clear that glowing reviews were not enough to prevent the Adams Motorbungalo from being a commercial failure.

WANTED: WEALTHY GASOLINE GYPSIES (1922)

Elon Jessup, author of *The Motor Camping Book*, had forecast the Motorbungalo's main problem in 1921:

> An outfit which possibly might be classed as being of the general "palace car" type but very much lighter in weight and hence more practicable than others is the Adams Motor Bungalo. The weight in this case is not greater than that of some of the trailers which I have described. It is considerably more luxuriant and by the same token costs three times as much.
>
> —From *The Motor Camping Book* by Elon Jessup (1921)

An auto camp in 1923 with many tents but few camping trailers. *Courtesy of Library of Congress (record 2016847862)*

Jessup was being generous. The luxury version of the Motorbungalo cost up to $1,200, whereas an Auto Kamp trailer, for example, could be purchased at the time for under $200. As hybrid camping trailers, Motorbungalos were perhaps too luxurious and costly for Ford Model T owners but not grand enough for the wealthy. This was particularly true in the postwar recession of the early 1920s, when jobs, disposable income, and leisure time were hard to come by.

Apart from price, a second issue affecting sales was how to get wealthy people to consider going camping instead of using hotels. Writing in *Vanity Fair* in 1923, George W. Sutton said,

> It is still true, of course, that the majority of people who seek health and recreation as gasoline gypsies are people of extremely moderate means, to whom motor camping spells a glorious freedom from the cares, restrictions and expenses of home life. But it is also being recognized that it is a sport equally enjoyable by people of means and culture and that it need not have any of the disagreeable elements of "roughing it."
>
> *Vanity Fair*, June 1923

This recognition among "people of means and culture" that camping could be comfortable came too late for Curtiss and Adams. Although over one million people took motor camping trips in 1922, most did so by packing a separate tent or attaching a fly sheet to their automobile. The newly established middle class of America was not yet ready to become "gasoline gypsies" on a large scale.

As a result, the initial sales spurt of the Motorbungalo petered out. It is possible that a May 1922 factory fire at Keuka Industries (which was supplying parts for the Motorbungalo) may have affected production. It is believed that only about a hundred Motorbungalos were sold. Adams followed Curtiss to Florida in January 1923, and the Adams Trailer Corporation was dissolved in 1924.

Lessons had been learned. Curtiss and Adams were selling two things: a new type of vehicle and a new lifestyle. Traditional automobile dealers would have had little comprehension and even less patience for either. Advertisements referred to how fast the Motorbungalo could travel, but why rush a vacation? A sales pitch that often focused on speed may have been irrelevant to most potential customers who were looking only for comfort.

While the production model flopped, Curtiss had been on the right track with the Curtiss Camp Car. It was, in effect, an early prototype of the Aerocar. It was large enough to give a feeling of indoor space and to incorporate many home comforts. But by making the production models smaller and retaining the partial use of canvas, they came to resemble other more affordable camping trailers. Its modest dimensions forced users to spend more time outside, so in effect they were still "roughing it."

From a design standpoint, the Curtiss Camp Car was revolutionary. Its fifth-wheel hitch was almost certainly the first to be used in an American recreational vehicle. A more sophisticated version of the hitch would be seen later in the Curtiss Aerocar. If the fledgling American recreational-vehicle market had not been ready for the Motorbungalo in 1920, Curtiss would have to wait for it to catch up.

He would have to wait until 1928.

CHAPTER 4

CURTISS IN FLORIDA (1920–30)

In some unknown manner Valentine Day slipped by without any manifestations. Perhaps it is the climate down here which helps to conceal these holidays or perhaps it is because every day seems like a holiday.

Street view of the Opa-locka Company Administration Building (1930), one of the many properties developed by Curtiss in Florida. *Courtesy of the Gleason Waite Romer Photographs Collection, from the Miami-Dade Public Library System*

The Aerocar was conceived and first built in Florida. Curtiss's activities and connections in Florida were instrumental in its birth.

INVESTING WITH A PURPOSE

Miami is to the Aerocar what Hammondsport is to Curtiss's motorcycles and planes. While the Aerocar may have been inspired by the Motorbungalo prototype built in New York, its birth eight years later was unexpected and took place in Florida.

We have learned something of the philosophy behind the Aerocar, but to understand where it fits into America's social and transport history, it helps to know something about Curtiss's activities in Florida in the last decade of his life, between 1920 and 1930.

Curtiss's final decade was one of the most active of his life. In southern Florida, he continued to develop ideas and initiatives in a far broader range of endeavors than before. These ranged from commercial aviation to new road vehicles and from community development to indigenous tourism, horticulture, and even movie-making. He pursued his hobbies of hunting, archery golf, and others and, perhaps for the first time since his youth, was able to relax.

Curtiss became a major real estate investor in southern Florida, but for Curtiss, real estate was a means to another end rather than an end in itself. Around 1920, Florida was packed in January but nearly empty in June. For the winter paradise to flourish year-round, Curtiss felt it needed more than just tourism. It would need a permanent and skilled population, affordable housing, new infrastructure, and a diverse industrial and agricultural base. So Curtiss became a community builder. He chose to do so at a time of intense national interest in Florida, and over a period that included significant economic and climatic shocks. He would be in for another bumpy flight.

Curtiss (*left*) relaxing on the beach with family and friends (date and place unknown)

A group of horseback riders in front of the Miami Biltmore Hotel (1931). *Courtesy of the Gleason Waite Romer Photographs Collection, from the Miami-Dade Public Library System*

THE FLORIDA REAL ESTATE BOOM (1920–25)

The end of World War I saw a confluence of circumstances that drew thousands of people from the North to spend their winters in sunny southern Florida. As early as 1876, the Biscayne Bay Company had described the area along the Miami River as "more healthful and free from disease than any other section of the Union." The region's popularity as a tourist destination grew in the 1880s thanks in the main to the railroad and hotels built by pioneering industrialist and developer Henry Flagler, who like Curtiss was from Hammondsport.

During and shortly after World War I, the European Riviera had become off-limits to wealthy American travelers. They needed an alternative. Florida seized on this opportunity by adding enticements to developers such as cheap land, low-cost finance, and low or zero taxes. With an extensive tourism advertising campaign in northern cities, improved road and rail links, and a relaxed approach toward Prohibition, it was a heady mix. What could possibly go wrong?

So in 1920 began a five-year "hullabaloo" in southern Florida. Fueled by real estate acquisition, development, and speculation that at its peak saw land being sold and resold several times in one day, Miami and its surrounds were transformed by new towns, cities, and infrastructure. The region saw rapid growth in tourism, leisure pursuits, aviation, and agriculture. It was also a period that saw dislocation of the Indigenous Seminoles, continuing overt racism, and significant environmental and wildlife destruction.

The Florida boom came to an end in late 1925, when road repairs blocked incoming building materials, tax inspectors found land to be overvalued, banks became cautious, land buyers became scarce, and real estate values fell as quickly as they had risen. Two hurricanes in September 1926 and September 1928 killed thousands and demolished or severely damaged tens of thousands of buildings. The hurricanes reminded both speculators and tourists that nature could not be easily tamed, and that living in Florida had a major downside. These events put a brake on the further significant development of Florida until the 1950s.

FLYING AND FLORIDA

In common with many other wealthy northern industrialists, Curtiss moved to Florida in part because of its climate. During winter, the warmth and sunshine of South Florida became an irresistible attraction for many northerners, especially those in poor health. But Florida held a second attraction for Curtiss. It was ideally located as a future center of aviation for flights from North to South America and from Cuba to the Caribbean. Curtiss felt that he could contribute to the growth of Florida as an airline stopover point and an aviation hub for central America. If the planes had wheels, Florida had plenty of land to create airports. If the planes had floats, Florida had suitable coastline and harbors.

Curtiss had known Florida since the day he sped along Ormond Beach in a motorcycle in 1907. In 1912, he established a flying school in Miami. When building his early aircraft, Curtiss soon came to realize that for this new industry to grow, it would require two things: pilots and somewhere to take off and land. Curtiss was instrumental in creating both through his flying schools and airfields across the country.

During the 1910s and after some initial skepticism, Curtiss grew increasingly confident about the prospects

An oversized sign of Seminole Indian Jack Tigertail welcoming investors to Hialeah (1921, *left*), and prospective land purchasers at Triangle Park, Hialeah (1921, *right*). *Both images courtesy of Hialeah Public Libraries, Hialeah History Photography Collection*

for commercial airline travel. But if one day, long-distance passenger flights were to become commonplace, they would still need stopping-off points along the way. Curtiss was particularly interested in the commercial possibilities of seaplanes because they didn't require airfields, although in reality they would prove to be logistically more challenging to operate commercially than land-based operations.

Curtiss first became involved in Florida real estate when he began looking for a site for an airfield there in 1917, well away from the costly and increasingly crowded seafronts and city centers. He went looking in southern Florida's "back lots." As he bought and sold land there over the next few years, Curtiss would use some of his vehicles, both old and new, to transport goods and passengers and help build new communities.

THE CURTISS TRAILER BUS (1922)

An unusual Curtiss vehicle made its way from New York down to Florida shortly after Curtiss and his family moved to Florida in 1920. It was the Curtiss Trailer Bus. Although outwardly more mundane in design and purpose than some of Curtiss's other road vehicles, the Curtiss Trailer Bus was nevertheless an important stepping-stone in engineering terms between the Adams Motorbungalo camping trailer of 1920 and the Aerocar of 1928. It was a rare early example of a fifth-wheel passenger bus. It was designed by Curtiss and built by Curtiss mechanic John Thiel, to whom Curtiss, according to the *Miami News*, had sent "a pencil sketch and a few notes on how he wanted the trailer to be made, with no measurements and no specific design, other than the turntable principle."

The Curtiss Trailer Bus with a Dodge chassis (1922). The logo of manufacturer Adams Trailer Corporation is just visible behind the driver's seat.

The first Trailer Bus was built in 1922 by Adams Trailer Corporation of Garden Island in New York, since sales of the Adams Motorbungalo had slowed. It was delivered to Florida for a specific purpose; namely, transporting potential real estate investors between downtown Miami and Curtiss's first new community in Florida, Hialeah.

Using a semitrailer to carry passengers in 1922 was novel. At the time, the American haulage and passenger transport industries had a clear preference for trucks over semitrailers, so convincing them to try something new was not easy. "Is not the horse and wagon an example of a load being pulled instead of carried?" asked *the National Taxicab and Motorbus Journal* in January 1922 when attempting to persuade its readers of the benefits of the Trailer Bus. To prove how easy it was to pull a passenger trailer, Curtiss used an ordinary Dodge runabout as the tow vehicle. It was modified to reduce overall length and to enable the driver to collect fares from up to twenty-six passengers. The Trailer Bus used the same "turntable principle" that Curtiss had used in the Curtiss Camp Car of 1919, along with the Martin Rocking Fifth Wheel most likely used in the Camp Car.

On its initial journey from New York to Florida, the bus covered some 1,700 miles (2,736 km) in ten days. Thiel, the driver for the journey, commented later how well the bus dealt with the poor roads. Reviews noted its tight turning circle of 40 feet (12.2 m) and its comfort when carrying light loads.

To support growing interest in land ownership at Hialeah, a second, larger Trailer Bus was built in Florida in late 1922. It was built by Thiel and Carl Adams under the name of the Hialeah Coach Company. The second Trailer Bus had a ¾-ton Garford truck chassis, an increased seating capacity of forty people, and a specially designed swivel seat for the driver. Orders for four additional coaches were reportedly received from unspecified customers in April 1923.

A double-decker version of the bus was announced in February 1923, with a capacity of seventy-five passengers. As far as we know, only one was built, serving as tourist transport in Miami into the late 1920s and 1930s. In the overall design of the Trailer Bus, we see more early design clues to the Aerocar. It was a predecessor to a second Curtiss-designed passenger transport vehicle that would be launched in 1930, called the Aerocoach.

The Trailer Bus carrying prospective land purchasers at Hialeah near Miami (1921). *Courtesy of Hialeah Public Libraries, Hialeah History Photography Collection*

A double-decker tourist bus built by Aerocar (1934). *Courtesy of HistoryMiami Museum*

THE CURTISS-BRIGHT CITIES

Curtiss began spending a significant part of each winter in Florida from 1917 onward. In his search for suitable land to establish an airfield outside Miami, Curtiss was introduced by Miami Chamber of Commerce president (and later Miami mayor) E. G. Sewell to a dairy farmer named James Bright. Bright owned with his brother a 1,200-acre dairy farm in Dade County. Bright shared Curtiss's views on the potential for the cautious development of the back lots of southern Florida rather than more speculative waterfront or city-center investments. Curtiss felt that the Bright farm was ideally located for an airfield, but as they discussed the right type of grass for the field, Curtiss became interested in the agriculture of the region. It's possible that he considered it about as backward as the automobile industry of the time and ripe for some innovation.

The airfield was established in 1917. Because of the ongoing war in Europe, Curtiss immediately donated it to the US Navy as an air base. Following the end of World War I, Curtiss teamed up with Bright in 1919 to form the Curtiss-Bright Ranch Company (the word "Ranch" was later dropped), capitalized at $1 million.

Leisure facilities were key components of all Curtiss-Bright cities.

The start of the Derby at the Hialeah Park Race Track (1934). *Courtesy of the Gleason Waite Romer Photographs Collection, from the Miami-Dade Public Library System*

The Hotel Country Club at Country Club Estates (date unknown)

The Miss Miami beauty pageant at the Opa-Locka Pool (1927). *Courtesy of the Gleason Waite Romer Photographs Collection, from the Miami-Dade Public Library System*

A group of archery golfers at the start of a match at Hialeah Golf Club (1923). Conventional golfers played against archers, whose aim was to hit a coconut placed near the hole. *From left to right*: Mike Brady, Walter Andrews, Sam Huff, and Glenn Curtiss.

A second company with the same capitalization, called the Florida Ranch and Dairy Company, was set up to manage the dairy side of the partnership. Curtiss initially acquired 5,000 acres adjacent to the Bright ranch for future uses that included the development of communities that would provide farmers somewhere to live. It was reported that Bright persuaded Curtiss to become more ambitious in the community development area, suggesting that the partners establish new towns not just for farmers, but for anyone. Curtiss reportedly agreed, and more land was purchased for future expansion. The Curtiss-Bright Company would eventually own 125,000 acres in the area, which would be used to create not only farming land, but also the planned cities of Hialeah, Country Club Estates, and Opa-locka, known collectively as the Curtiss-Bright cities.

Curtiss's early vision for his newly acquired land was the creation of communities incorporating industry, commerce, and agriculture with affordable housing for local workers, as well as family and friends who might choose to relocate to Florida from the North. Early land sales went well, and tent-and-hut villages were established in the area as eager new residents awaited completion of their permanent homes. Over time, the villages became towns and towns became cities. As revenues flowed from land sales in the early 1920s, Curtiss's plans became grander.

The sections of Curtiss-Bright lands originally selected for development were to become the three towns of Hialeah (1921), Country Club Estates (1924, renamed Miami Springs in 1930), and Opa-locka (1926), all north of the current Miami International Airport.

HIALEAH (1921)

Hialeah was established on Everglades swamplands drained by Bright. The town's name was derived from the Seminole word meaning "pretty prairie" or "prettiest pearl in the heap." It was marketed as "the Gateway to the Everglades."

When complete, Hialeah encompassed among other things a large mansion for Curtiss, Bright's ranch and house, Willy Willy's Indian Village (an Indigenous tourism village named after the local Seminole chief), the Curtiss Aviation Field, Hialeah Park (horse racing), a greyhound racing track, a jai alai fronton (jai alai was the national racket sport of Cuba, played on a court called a fronton), a movie studio, a sugar plant, a soap factory, a power-generating plant, a water plant, a city hall, a post office, churches, and even a deer park. A Hialeah landmark was a large cutout of Jack Tigertail, a Miccosukee Indian pointing in the direction of the Hialeah real estate office. The Woman's Club of Hialeah was established by Lua Andrews Curtiss, Glenn's mother, who moved to Florida to live in Hialeah in 1922.

An important part of the housing policy at Hialeah was affordable housing, especially for people of color. This innovation would help local industry become established in an otherwise challenging area, and made home ownership possible for many first-time buyers.

Initial land sales were promising as investors were bussed into the area from Miami on Curtiss Trailer Buses. By 1925, Hialeah had twenty-five factories. But over time, the racetrack would attract gamblers and drinkers, with "Hialeah hooch" becoming a well-known local beverage. Despite Bright's drainage efforts, Hialeah was subject to regular flooding.

COUNTRY CLUB ESTATES (1924, RENAMED MIAMI SPRINGS IN 1930)

Country Club Estates was begun when Hialeah was reaching capacity. In a period when Florida real estate was often bought sight unseen by northern investors, property had to stand out visually. It needed a theme. For the architectural theme of Country Club Estates, Curtiss chose the Pueblo-Revival style. It was a small, mostly residential development that became popular first with railroad and later airline workers. Its centerpiece was a golf course, which was the first in Florida to allow Black people to play, and a hotel. The Hotel Country Club, completed by Curtiss with unfortunate timing in 1927, was badly affected by the local economic depression that followed the Florida real estate bust and 1926 hurricane. It was sold in 1930 to Dr. John Harvey Kellogg of breakfast cereal fame at a nominal sum for use as a health sanitarium. Country Club Estates was renamed Miami Springs in 1930 after the natural spring found under its golf course. Curtiss deeded these water rights to the City of Miami. Curtiss built a new home for himself in Country Club Estates, as well as two homes for his mother. G. Carl Adams also lived there and later became mayor of the city from 1930 until 1942.

OPA-LOCKA (1926)

Opa-locka (also "Opalocka") was the third community to be developed by Curtiss. Once again, an enticing architectural style was deemed necessary. The theme for Opa-locka chosen by Curtiss and delivered by New York–based architect Bernhardt Müller was the Moorish architecture similar to that seen in the Arabian Nights fables. Sometimes called "the Baghdad of Dade County," Opa-locka's domes, minarets, parapets, and outside staircases were standout features of the town's eighty-six main buildings, with the Opa-locka Administration Building the most prominent example.

Despite architecture that many today would judge garish, Opa-locka's planning and layout were more considered than Hialeah. Greater controls were placed on buyers. Its roads were built on a pattern of concentric circles rather than a grid. A golf course, an archery club, a bank, and a small zoo were built among the private residences. Importantly, the town had a railroad station. The Seaboard Air Line railroad stopped at Opa-locka, considerably increasing the appeal of the city.

Opa-locka would also suffer in the 1926 hurricane but flooded less frequently than Hialeah. It was the location for the first Aerocar plant at 1170 Sharazad, although this building was constructed some time before the plant commenced operations there in 1928. Early Aerocars were frequently photographed in front of the striking Opa-locka Administration Building.

Curtiss family houses in Florida

Glenn Curtiss's first residence in Florida at Deer Park, Hialeah (1921). Curtiss lived here from 1921 to 1925.

"Dar-Err-Aha," Curtiss's second residence, at 500 Deer Run, Country Club Estates (ca. 1925). Curtiss lived here from 1925 until his death in 1930.

The flooded home of Curtiss's mother, Lua Andrews Curtiss, in Hialeah (1922). Curtiss built a second home for his mother at Country Club Estates in 1925. *Courtesy of Hialeah Public Libraries, Hialeah History Photography Collection*

The home of G. Carl Adams in Hialeah (1924). Adams also moved to Country Club Estates in 1925.

MIAMI STUDIOS (1922)

Left: A movie set inside Miami Studios, Hialeah (1922). *Courtesy of Hialeah Public Libraries, Hialeah History Photography Collection*

Above: An external view of Miami Studios (1925). *Courtesy of the Gleason Waite Romer Photographs Collection, from the Miami-Dade Public Library System*

The president of the Miami Chamber of Commerce, E. G. Sewell, was a strong believer in the prospects for sun-drenched southern Florida to rival California and New York as a hub for making silent movies. In 1922, Curtiss agreed to allocate land at Hialeah for the construction of a movie studio and film-processing laboratory. He saw this as an opportunity to create a new industry at Hialeah. Miami Studios, as it was to be called, was built in record time and opened to great fanfare with the "Cinemiami Ball" in March 1922. Curtiss was appointed president of Miami Studios, Inc.

Curtiss was interested in photography from an early age and would have been fascinated by the technology behind moviemaking. He built not only two film stages but also a processing laboratory and a dedicated power plant. He placed at the studio's disposal a mobile lighting unit for location shots (carried on a trailer) and arranged for a large portable wind machine to be brought down from Hammondsport for use in the special-effects department.

The studio had early teething problems, such as local flooding and the lack of experience of local production crews, including carpenters who were not used to making fast and flimsy film sets.

These problems were soon ironed out, however. Between 1922 and 1924, a number of silent films were made at the studios, including *Isle of Doubt* (1922), *Swamp Demon* (1922), *Outlaws of the Sea* (1922), *The Glade Legend* (1922), *The White Rose* (1923), *Where the Pavement Ends* (1923), *Ramshackle House* (1924), *Icebound* (1924), and *Another Scandal* (1924).

Two major handicaps kept the life of Miami Studios short. The first was Florida's terrain. The lack of any nearby mountains, deserts, forests, or steams limited scope for Florida-made movie plots. The second was the end of the Florida real estate boom in 1925, depressing values of land and buildings in the area and making Florida film financing challenging. The Great Miami Hurricane of 1926 was the final straw for the studios, which were extensively damaged. By 1927, the studios had ceased to exist.

CURTISS HOBBIES IN FLORIDA

Before Curtiss's financial troubles started to grow in the late 1920s, he seemed more relaxed in his early Florida days than at any time in his life. He was no longer under the intense pressures of manufacturing aircraft to schedules and budgets. He had escaped the harsh New York winters and had more time to indulge in some of his hobbies.

The sport that Curtiss seemed to enjoy most during the 1920s was archery. The logo of the Aerocar Company of Detroit would later include an arrow, no doubt reflecting the speed and aerodynamic properties of the Aerocar. A particularly novel type of archery that Curtiss played was archery golf. This is where coconuts would be placed near a golf hole, and archers would aim to hit the coconuts in fewer shots than conventional golfers playing alongside with golf clubs and balls.

Curtiss was also a keen hunter. He is reported to have hunted everything from birds to bears and from deer to alligators. One of the first structures he built at Hialeah was a modest hunting lodge, which also served as a hideout whenever Curtiss wanted to retreat from social engagements at home. It was later turned into a school.

> Got thirty doves in about an hour day after you left[;] also located excellent duck shooting.
>
> —Telegram from Curtiss to C. M. Keys, December 14, 1927

Most of Curtiss's camping activities were undertaken as part of hunting trips. Curtiss went to Europe in 1927 and spent some time at a hunting lodge in Scotland. Other sports played or promoted by Curtiss on his properties were trap shooting, lawn bowls, and jai alai. Curtiss also had a small zoo and a deer park at Opa-locka, both intended as tourist attractions.

Curtiss was not the only person to pursue leisure in Florida. Carl Adams had an interesting hobby of his own. Alongside his early bus-building activities, Adams took up with mechanic John Thiel the unusual sport of auto polo in 1924. Using specially modified autos instead of horses to play polo was one of the many fads of the 1920s. It was not for the faint-hearted, however, and never gained a broad following.

Curtiss the hunter (dates and places unknown)

Curtiss (*left*) with a young captive alligator. Curtiss would generally catch and relocate alligators away from human habitation rather than hunt them for sport.

Curtiss after a successful duck shoot

Curtiss with bow and arrow

Curtiss on a hunting trip with guide and dog in an early Aerocar

The hunting lodge built by Curtiss at Hialeah. It later served as the headquarters for Miami Gun Club and Archery Golf Club and was used as an emergency school after the 1926 hurricane (date unknown). *Courtesy of the Gleason Waite Romer Photographs Collection, from the Miami-Dade Public Library System*

Two auto-polo enthusiasts, Bob Millard and Dr. Lamb, with a vehicle converted for auto-polo use at Curtiss Aviation Field (1922). *Courtesy of Hialeah Public Libraries, Hialeah History Photography Collection*

GREEN-FINGERED GLENN

Underlying Curtiss's Florida investment philosophy was the importance of creating self-sustaining communities. He felt that Florida needed farmers as well as millionaires, and food as well as frivolity. To manage the region's challenging soils, irrigation, and climate, Curtiss became a horticultural innovator.

In 1919, Curtiss tried breeding goats for the production of angora wool. In 1920, he brought five grape-growing families from Hammondsport to try to grow grapes in the area. Curtiss's grandmother had owned a vineyard in Hammondsport, and Curtiss spent many happy days there in his youth, collecting grapes and rushing them to market at harvest time in a tractor and trailer. In 1924, Curtiss experimented with aerial seed sowing from a plane, using a so-called "aeroplanter," a hopper specially designed to drop grass seed from a plane. These and a number of other horticultural trials failed due to extreme weather and poor irrigation, but Curtiss was experimenting on the farm as he did in his workshop and would learn as much from failure as success.

Curtiss was an outspoken critic of state drainage systems and their inability to deal with the inundations often seen in the area. He pushed for a better road network through the Everglades for firefighting access after a topsoil blaze in 1925. After the hurricane of September 1926, Curtiss and Bright not only donated land to local farmers to help them get back on their feet, but offered to plow their land using the Curtiss-Bright tractor fleet. In 1927, Curtiss used 250 small stoves to prevent frost damage to his papayas and bananas.

The extent of the farming challenge on Curtiss's initial land choice can be seen from this desolate-looking and flood-prone part of the Curtiss-Bright dairy farm near Hialeah (1921). Farmers' houses can be seen in the background. *Courtesy of Hialeah Public Libraries, Hialeah History Photography Collection*

Curtiss with a papaya tree (date and place unknown)

One of the first products to be used in the Aerocar was Celotex, a locally produced insulating material made from sugar cane fiber that was used on the walls of early Aerocars. A decade before Henry Ford's love affair with soybeans, Curtiss was seeking to align agriculture with industry.

BRIGHTON (1925)

Frustrated with regular flooding events on its land at Hialeah, in 1925 the Curtiss-Bright Company purchased 28,000 acres (later increased to 40,000 acres) in order to establish a ranch midway between Miami and Tampa. Given the name of Brighton after James Bright, the ranch was located on high ground northwest of Lake Okeechobee. It was less prone to flooding, although just to be on the safe side, the company dug 27 miles (43 km) of drainage canals in the area. Brighton was closer to the other major cities of northern and western Florida. Local Seminole tribe members were given assurances that they could continue to live, work, and hunt freely on the newly acquired land. Brighton had at its core a number of privately owned "truck farms" collectively called the Brighton Valley Farms. Known also as market gardening, truck farming took place on the fringes of an urban center where fresh produce was trucked daily into town for sale to local hotels and stores.

The Palm Circle Dairy near Brighton (1926)

As well as the ranch that included over 1,500 head of cattle managed by James Bright, early agricultural trials at Brighton Valley included poultry, rice, cotton, bananas, watermelon, and even "tearless onions." The experimental farm became a destination for farmers nationwide to see how Curtiss-Bright's successes (and failures) could be applied to them. Curtiss and Bright offered significant financial incentives and low-cost housing to entice farmers to relocate permanently to Brighton, where, according to its advertising, "soil, toil and sunshine can be combined to produce profits in every season."

A hotel called the Palm Circle Inn was opened in Brighton in March 1926, where visitors were invited to pause on their journey south. They could also choose somewhat incongruously between visiting a zoo or taking a guided hunting trip. One report suggests that the Brighton Zoo kept ten Galápagos Islands tortoises for breeding purposes, due to their endangered status in their native country. Later, in 1929, a dude ranch was established where city dwellers could experience rodeo and other ranching activities.

Brighton used a cooperative farming model. Free farming advice was provided to all landholders. Farmers would sell their produce at uniform prices to Curtiss-Bright, who would then transport and sell the produce to stores in Miami and other cities, including local Piggly Wiggly stores, America's first self-service groceries. To help them get their produce to market, a 150-mile (241 km) trip in the case of Miami, Curtiss provided Brighton farmers with an unusual form of transportation—a fifth-wheeled semitrailer that in shape and size was not unlike the Motorbungalo prototype. We shall meet this semitrailer in the next chapter.

CONSOLIDATION AND CONCERN (1927)

The business consequence of creating a disconnected, sprawling portfolio of moneymaking and not-for-profit schemes in such a short time by a man who was self-admittedly "not good with numbers" was mild financial chaos. When overlaid by economic and climatic disruption, the mild chaos became severe. In 1927, financier C. M. Keys once again came to Curtiss's financial rescue by consolidating Curtiss's Florida companies into a new entity called Glenn H. Curtiss Properties, Inc. As an astute investor, Keys felt that the end of the Florida real estate boom created a good investment opportunity, so he invested personally in the new company. But Curtiss's businesses needed strong financial control at the local level rather than helicopter investors, and the reorganization was neither early nor extensive enough to insulate Curtiss's businesses from their next financial shock—the Wall Street crash of October 1929. Curtiss's business partners and financiers suddenly had much-bigger problems on their plate, leaving tracts of land in southern Florida low on their list of priorities. As Curtiss entered what would be the final year of his life in 1930, his Florida investments were worth a fraction of their value at the height of the boom.

CARL G. FISHER

Carl G. Fisher (1916). *Courtesy of the Gleason Waite Romer Photographs Collection, from the Miami-Dade Public Library System*

The Miami Beach Regatta in front of the Fisher-owned Flamingo Hotel (1928). *Courtesy of the Gleason Waite Romer Photographs Collection, from the Miami-Dade Public Library System*

Carl G. Fisher was from Indianapolis. He had made his fortune as co-owner of Prest-O-Lite, a company making acetylene automobile headlamps in the early 1900s under a French patent. The company was sold to Union Carbide in 1911 for $9 million, making Fisher a wealthy man. Fisher also conceived and coinvested in the Indianapolis Raceway in 1909 and helped secure financing for the Lincoln and Dixie Highways. He loved automobiles, racing them in his early days and later owning an automobile dealership. Although he didn't have the engineering skills to build automobiles himself, he never shied away from telling others how they should be improved, particularly in the area of comfort.

Fisher had been drawn to Florida in part because of its real estate opportunities, but also because of a climate that he felt would help his breathing difficulties. Fisher was known to everyone in Miami as the creator of Miami Beach, building this new Miami suburb from swampland. He dredged waterways and removed mangroves to make way for new hotels, residences, golf courses, and polo fields. Fisher was a consummate promoter. Among his publicity stunts were transporting an automobile (with the engine removed) by hot-air balloon, dropping an automobile (with tire pressure reduced) from a high building, and using an elephant as a golf tee. He was well connected in the automobile industry.

Curtiss had known Fisher as far back as 1909, when Fisher invited him to consider holding an aviation show and establishing an air school in Indianapolis. Fisher was one of several people to encourage Curtiss and other wealthy northerners to consider investing in Florida. His sales pitch was hard to resist.

Fisher's role in the history of the Aerocar is significant because, thanks to a chance ride with Curtiss, he lifted a farming truck from obscurity into one of the best-known travel trailers ever built. If Curtiss was the Aerocar's designer and engineer, Carl G. Fisher was its talent spotter and salesman.

A RIDE THAT CHANGED EVERYTHING (1928)

For Curtiss, the period between the launch of the Adams Motorbungalo in 1922 and the introduction of the Aerocar in 1928 was one of building communities rather than machines. During this time, he would continue to support the growth of the commercial airline industry, but more as an ambassador than an inventor. The Curtiss-Bright cities as well as rural Brighton had generated new industries and employment but were indirect casualties of the real estate collapse of 1925. Curtiss's horticultural experiments were of great local benefit but time consuming and generated little if any income.

By the mid-1920s, Curtiss had become preoccupied with his basket of Florida problems. But he persevered in the hope that what he sowed today he would reap tomorrow. It's more than likely that he would have spent the rest of his days battling against the markets and the elements in southern Florida had it not been for a ride that Curtiss gave in his farming semitrailer in early 1928 to one Carl G. Fisher.

CHAPTER 5

THE CURTISS AEROCAR (1927–30)

You know, Glenn is a very conservative man and very much more on mechanics and philosophy than he is a salesman. I don't believe Glenn could sell gold dollars for ninety-five cents apiece, but he could certainly make gold dollars as good as the mint could make them if he wanted to do so.

—Carl G. Fisher, April 1928

A Curtiss Aerocar with a 1929 Hudson in front of the Opa-locka Company Administration Building (ca. 1930)

A Brighton Valley Farms trailer towed by a Ford Model A on a beach in southern Florida. James Bright is in the center; the other people are unknown. (ca. 1928).

THE BRIGHTON PROTOTYPES

In April 1928, Carl G. Fisher made a two-day tour of central Florida. As part of that tour, he visited Brighton and took a ride with Curtiss in a trailer. It was the one used to transport Brighton Valley Farms produce to customers in Miami. According to a later report by Fisher of this trip, the trailer was "necessarily crude and not a great deal of attention paid to lines" but was in other respects, in Fisher's view, sensational.

According to Fisher, the trailer's most remarkable features were its low cost, light weight, and supremely comfortable ride, the latter achieved mainly through the use of a hitch that consisted of an aircraft tire placed horizontally at the rear of the tow vehicle, which absorbed road shocks and vibrations.

From the evidence that remains today, there is no suggestion that prior to this trip with Fisher, Curtiss had any plans for his trailer beyond its continued use in truck farming and as a promotional vehicle for his real estate and agricultural ventures. He was clearly preoccupied. It is to Fisher's credit that this trailer would before long become known across the country as the Aerocar.

Fisher's nose for a new business opportunity sniffed something stronger in this trailer than fruit and vegetables. He was totally overcome by the experience and immediately began to think of new applications for the trailer. In a letter to C. M. Keys, he gushed,

> I think Glenn has another little gold mine in that Trailer of his . . . there are two or three big points that I believe if covered by patents would give Glenn a strangle hold on the light trailer business in this country. Certainly, nobody could make a trailer as cheap as Glenn is making, for no trailer I have ever seen can compare in easy riding, light weight[,] and the adjustment of a noiseless coupler or universal joint, which is really air compression in rubber.

A trailer in front of the Opa-locka Company Administration Building. Following Fisher's intervention, the trailer was quickly refined by Curtiss engineers into the first Aerocars. (ca. 1928).

> This is one of the cleverest engineering features in universals I have ever seen, and there is no reason in the world why this universal cannot be used on street car connections, also on light railway connections—possibly not in this particular form but in an adjustable form that would be practical for the intended work."
>
> —Letter from Carl G. Fisher to C. M. Keys, April 28, 1928

Fisher could not believe that Curtiss was planning nothing more ambitious for his trailer than to serve as a rolling billboard advertising Florida land sales to farmers in Wisconsin. In 1928, Fisher, who was facing his own financial problems following the decline in value of his Miami Beach real estate and slow sales of his next development at Montauk on Long Island, was probably happy to have a diversion away from real estate into his beloved area of automobiles.

Fisher and Curtiss began forming a plan to refine, patent, and manufacture the trailer. The Fisher publicity machine quickly swung into action. This first article dealing with an Aerocar prototype is worth quoting in full. Under the heading "Miamians Contrive Automobile Trailer. Glenn H. Curtiss and Carl G. Fisher May Build Manufacturing Plant in Miami," the *Miami Herald* wrote,

> Glenn H. Curtiss, Carl G. Fisher and a few friends are interested in a new type of trailer on which they have applied for a patent and which they will construct in a factory in Miami. A few of the vehicles have already been built and have been praised by all who have seen them.

In devising the vehicle, which is of the semi-trailer type as it runs on two rear wheels with the front supported by the rear of the tow-car, Mr. Curtiss used the principle on which the fuselage of an airplane is built. That is, there is no chassis or heavy metal frame to support the structure. Being constructed like an airplane it is self-supporting and is covered with celotex which is light and also acts as insulation to keep out heat, cold and noise.

Mr. Curtiss constructed the first one with the intention of installing in it a display of fruits and vegetables grown in Florida together with samples of the soils in the Curtiss-Bright properties and sending it for a tour of the northern states. Mr. Fisher saw it and was so impressed with its merit that he ordered one for himself and became financially interested in its development.

The trailer can be used for the transportation of freight, it can be fitted with windows and seats for use as a bus or it can be equipped like a small bungalow for touring. Gen. Robert H. Tyndall has also called attention to the possibilities of the vehicle for military use as an ambulance. Its advantages are its light weight and economy in construction and operation together with its easy-riding characteristics.

Mr. Curtiss has devised a folding berth which is supported by heavy rubber bands which entirely eliminate all vibration even when the car is driven at high speed. He has tested the trailer and found that in actual use it is better even than he had regarded as possible. The "fifth wheel" on which the front end of the trailer rests also embodies new principles which will permit the trailer and tow-car to turn in but little more space than the trailer occupies.

—*Miami Herald*, May 9, 1928

The article was accompanied by a photograph of the trailer crossing a bridge, painted white and bearing the words "Brighton Valley Farms" on its side.

So here was a vehicle, closely resembling the Curtiss Camp Car of 1919, designed apparently by Curtiss to transport fruit and vegetables. We learn that "a few of the vehicles have already been built," but how many is not revealed. In the same way that the Curtiss Camp Car had been the prototype of the Adams Motorbungalo, the Brighton trailers built in about 1927 were the prototypes of the Aerocar.

AEROCAR NUMBER 1

The name of Curtiss's new vehicle was first revealed in May 1928, when, according to the *Miami Herald* of May 22, 1928, Curtiss took a party of local businessmen and archers to Key Largo in an "aero-car." They were there to study a growth of snakewood, which was used to make archery bows. The vehicle was backed up against the water and used as a portable bathhouse. Later, in August 1928, a photo of the vehicle appeared in the press.

It was described by the newspaper as "the first Aerocar built at the Opa-locka plant by the Aerocar Company of Opa-locka." It was to be driven by Curtiss Properties employee J.A. Daly on a 7,000-to-8,000-mile trip lasting three to four months to promote investment opportunities in Curtiss properties in Florida to northern residents. It was a rather garish rolling billboard that may have retained some basic accommodation inside for Daly. The word "CAR" on the front left wing suggests that the word "AERO" may have been on the obscured right wing. The word "Aerocar" was also used on the side of the trailer. Because it is the first-known

The first trailer to be given the Aerocar name (1928). It was used as a promotional vehicle for Curtiss Properties. The first primitive Aerocar logo of a tire and wing is visible on the side, and the text includes references to some of the first local products used in the Aerocar's manufacture, Celotex and Rub-Ros. *Courtesy of Curtiss Mansion, Inc.*

vehicle to be clearly described and illustrated as an Aerocar, we shall call it Aerocar Number 1.

Interestingly, the external advertising included the names of two local companies whose products were used in the making of the Aerocar: Rub-Ros and Celotex. Rub-Ros was a patented sealant made from rubber and rosin. Developed by L. Francis, it used old tire inner tubes and other used automotive rubber components to make a rustproof coating. Celotex was a lightweight, weatherproof fiber board made from sugar cane and used on the sidewalls of the Aerocar. Common to both products was a focus on the use of recycled or waste materials, something that Curtiss would have been keen to promote.

Between the appearance of these two photos in May and August 1928, Fisher set his own marketing machine into motion, writing in glowing terms about the trailer to C. M. Keys as well as Roy Chapin and Howard Coffin, both of Hudson Motor Company. Thanks to his background at Prest-O-Lite and in fundraising for the Lincoln Highway, Fisher was well connected with the "gasoline aristocracy" of Detroit and would use these connections to stimulate interest in the new trailer. Fisher suggested well-known Miami boatbuilder Gar Wood as a possible manufacturer of the trailer and offered tips on alternative uses and how to protect the patent. Most importantly, he agreed to buy four trailers for his new development at Montauk and offered to buy a one-third interest in the patent. In Fisher, Curtiss had found an investor, customer, and marketing manager in one.

Curtiss's reaction to Fisher's enthusiasm was typically community minded. In a letter to Keys in May 1928, he wrote,

> I had an interesting talk with Howard Coffin and Carl Fisher at the Yacht Club. Mr. Coffin wants one of the trailers now and sees an opportunity for quite a number. Mr. Fisher wants several and we are building one up to get his approval.
>
> The best thing about it is a little industry for this community right now. As a matter of fact, I had rather build them here without profit than somewhere else at a profit, as the benefit to the community from the industry would more than offset the loss of profit.
>
> I talked with Mr. Fisher about taking an interest in the Brighton development, which he agreed to do. If he will spend a little money up there it will compensate for his third interest in the trailer patents.
>
> After you talk with Mr. Fisher let me know what your suggestions are as to how to proceed. In the meantime, I will go right ahead with the patent and the construction of the first few jobs as personal matter."
>
> —Letter from G. H. Curtiss to C. M. Keys, May 7, 1928

Curtiss's view of the Aerocar "job" as a community project would no doubt have exasperated hard-nosed money men such as Keys and Fisher, but this was simply Curtiss's own brand of welfare capitalism at work. Curtiss declined Fisher's suggestion to have the Aerocar built by a local boatbuilder and instead established a plant in an existing industrial building at Opa-locka run by people he knew had the necessary skills to produce a lightweight trailer—a few of his engineers from his early flying days under the leadership of early Curtiss pilot and engineer Hugh Robinson.

Although Curtiss and Fisher were very different people, they were both optimists by nature and excited by new ideas. Fisher's broader vision for the trailer, his network of motoring enthusiasts, and encouragement of Curtiss through advance orders were important catalysts to the birth of the Aerocar. As long as the manufacturing could take place locally, Curtiss allowed Fisher to introduce it to potential investors and create some "ballyhoo" around the concept.

THE AEROCAR CORPORATION

In June 1928, Aerocar's corporate structure was announced. Manufacturing was to commence at Opa-locka immediately and at other plants in due course. The Aerocar Corporation was established in Delaware to hold existing and future Aerocar patents transferred to the company by Curtiss and to issue licenses to manufacturers in return for modest royalties. Royalty payments began at $50 per vehicle for the first one hundred manufactured, and fell to $25 for over a thousand vehicles.

Aircraft-inspired construction of the Aerocar, incorporating laminated wooden struts and piano wire tightened with bronze turnbuckles (1933)

The Aerocar's rear suspension consisted of softly sprung leaf springs (1933).

The lack of a traditional chassis reduced weight and created internal space for a wide range of uses (1933).

Its initial stockholders were Glenn H. Curtiss; Howard E. Coffin and Roy D. Chapin of the Hudson Motor Car Company; W. O. Briggs of Briggs Manufacturing Company; Carl G. Fisher; financiers C. M. Keys and James A. Willson; and lawyer Chester W. Cuthell.

The Curtiss Aerocar Company of Florida was established as the first licensed manufacturer (replacing the short-lived Aerocar Corporation of Florida), operating under Curtiss's supervision. H. Sayre Wheeler was appointed president and general manager of the Florida operations, G. Carl Adams was vice president and treasurer, and Hugh Robinson became the plant superintendent.

There were to be five Aerocar designs initially: a commercial car, a school bus, a camp car, a touring car, and a private road car. In true Carl G. Fisher style and with perfect timing, Fisher publicized his order by telegraph for five Aerocars for his new leisure development at Montauk, Long Island, in July 1928. The Aerocar rollercoaster had left the station.

THE AEROCAR'S CONSTRUCTION

In 1930, Curtiss acquaintance Colonel James Prentice gives us a clear and independent description of the Aerocar's construction:

> The trailer was supported at the front by this pneumatic coupler, while at the rear end he had two fore and aft leaf springs of conventional automobile design. The two rear springs were secured to a tubular axle which terminated in roller bearings in regular pneumatic automobile wheels. The springs were slightly tilted to absorb road thrusts fore and aft.
>
> The framework of the trailer was mostly of laminated wood longitudinals with solid struts, vertical and diagonals of airplane wire secured to metal butts[,] and angle plates the same as in an airplane fuselage. The trailer was streamlined by tapering the longitudinals to meet near the hauling pintle, while at the rear end there was secured a streamlined device. Streamlining was found to be very important. Without it the trailer was hard on the mobile engine[,] but with it the fuel bill was greatly reduced and the towing auto could be driven at greater speeds than without the trailer since the trailer stopped swaying and skidding and steadied the auto down.
>
> Thin panels of material such as celotex [*sic*] was pasted to the framework in two separated layers which were attractively covered with imitation leather or duco.
>
> —Colonel James Prentice, "The Influence of Aircraft Design on the Trend of Motor Vehicle Construction," *U.S. Air Services*, December 1930

Prentice succinctly describes the "stick and wire" type of aircraft fuselage construction carried across to the Aerocar, which Curtiss and his engineers were familiar with from their flying days. It was strong, lightweight, and adjustable. The struts and longerons were made from airplane spruce with bent and bowed members, including roof ribs, laminated from elm and spruce. All surfaces were coated with lead and oil paint for protection. The standard airplane brace wires were of heavy nickel steel, and the turnbuckle barrels were made of bronze.

The Aerocar's rear wheel housings were lined with rubber. Although the rear suspension used conventional automobile leaf springs, they were hung using rubber shackles and were later assisted by hydraulic shock absorbers. In order to facilitate towing speeds of up to 70 mph (113 km/h), the Aerocar was streamlined from front to back with a "bird-beak" hitch, V-shaped front, sloping fabric roof, and curved rear.

The evolution of the Aero Coupler

The aircraft tire of early Aero Couplers was encased in a wooden frame (ca. 1928).

The Aero Coupler was mounted above or in front of the rear axle of the Aerocar tow vehicle, usually a coupe to facilitate installation, hitching up, and turning (ca. 1928).

Later versions of the Aero Coupler had the aircraft tire encased in aluminum (ca. 1932).

A V-shaped fairing was later added to the hitch pin to aid streamlining (ca. 1932).

Due in part to Fisher's vision for Curtiss's new trailer as having multiple uses, the design of the Aerocar was flexible from the outset. It needed to accommodate campers, businesspeople, goods, and even animals. Passenger versions had large, wind-down windows complete with roller shades. A range of internal options included wicker or fabric armchairs, drop-down or convertible beds, cooking facilities, an icebox, a washroom, a toilet, and extensive storage. Walls and ceiling were lined with washable imitation leather, and floors were covered in linoleum.

In later years, many modifications were made to the Aerocar's design. Models manufactured in Detroit had significant differences from those made in Florida, primarily to reduce manufacturing costs so that Detroit Aerocars could be sold at a lower price. The Opa-locka factory was initially used for custom-built models, while Detroit was the mass producer. In later years, as trailer competition intensified, Opa-locka would switch to building standard models.

THE AERO COUPLER

> Glenn Curtiss has the greatest trailer that was ever made in America. It is the cheapest thing built, it is absolutely noiseless. It has a coupler joint that is simple as an old shoe but it is perfectly wonderful.
>
> —Letter from Carl Fisher to Roy D. Chapin, chairman of the Hudson Motor Car Co., April 30, 1928

The Aero Coupler (sometimes "Aerocoupler") was the brand name given to a Curtiss invention described in its 1928 patent as a "flexible coupling for vehicular structures." Of all the innovations found on the Aerocar, the underestimated Aero Coupler was the most radical. It made a semitrailer safe and comfortable over a wide range of terrain and at speeds hitherto undreamed of for a vehicle of this kind. It allowed the Aerocar salesmen of the day to claim with some justification that the trailer was riding on air.

The Aerocar was designed from the outset to carry passengers while in motion. This is something that is today illegal in most countries on safety grounds, but in the US at that time (and in a number of US states still today) was both legal and commonplace.

To reduce the transfer of vibration and shock from the road surface to passengers in the trailer, Curtiss designed a new form of fifth-wheel hitch based on an aircraft wheel. The inflated wheel was embedded into a wooden (later aluminum) frame that was horizontally attached above or slightly in front of the tractor vehicle's rear axle. The gooseneck front of the trailer contained a fixed vertical pin that dropped into the center of the wheel and locked into place. Rubber discs and a spring sat under the center of the aircraft wheel. As the vehicle moved, the tire of the aircraft wheel absorbed the horizontal shocks and the rubber and spring absorbed the vertical shocks, creating a smooth ride for passengers.

As well as comfort there was a more important benefit to using a fifth-wheel design for the Aerocar—safety. Because the Aerocar was capable of traveling at speeds far higher than previously seen for any passenger vehicle of the period, trailer stability was a significant issue. Curtiss may not have been an expert on towing dynamics, but he would have known from observations that a fifth-wheel hitch reduced trailer sway at speed. The Aero Coupler was the first fifth-wheel hitch to be used in a production recreational vehicle in the US, and one of the safest trailer hitches of any period.

The second Aerocar sets out on a trip from Miami to New York, towed by a Hudson coupe (1928).

AEROCAR NUMBER 2

On August 5, 1928, what we shall call Aerocar Number 2 set out from Miami to New York with an important purpose. It was longer than Aerocar Number 1 and far more elegant. It had a sloping roof, windows with rounded corners, no wheel arches, and more streamlined front and rear ends complete with a bird-beak hitch.

The Brighton prototypes and Aerocar Number 1 had lost the external drop-down beds of the Curtiss Camp Car in favor of "a folding berth supported by heavy rubber bands." In Aerocar Number 2, this concept was extended with fixed sides and internal beds. The fixed sides improved both insulation and streamlining and, from a marketing viewpoint, distanced the Aerocar from any notions among potential customers of "roughing it." The Aerocar had been transformed from a camping trailer to a full-sized, solid-walled travel trailer. In the process, it had become the first known such vehicle to be manufactured in America.

The side of the trailer still contained some advertising (with the signwriter still not having made up his mind whether "Aerocar" was one word or two), but for the first time in public it was clearly proclaimed as the Curtiss Aerocar.

Aerocar Number 2 had already made its first trip in June 1928. Royal Canadian Air Force pilot John Fields had completed a 1,300-mile (2,092 km) round trip to Brunswick in Georgia to allow it to be assessed by Howard Coffin of the Hudson Motor Car Company, who planned to use Aerocars in his development north of St. Simon's Island. This Fisher-inspired initiative paid dividends, since Coffin was impressed and placed an order.

The second, much-longer trip to New York and Detroit in August was to demonstrate the Aerocar to northern automobile manufacturers. Although Opa-locka would be retained as the main Aerocar plant, Curtiss had been advised by Fisher and others to license manufacturing of the Aerocar to one or more established automobile makers. This was partly to allow manufacturers to come up with their own Aerocar designs and partly to back up Curtiss's patent. If the patent was accepted by well-known automakers, it

The Curtiss Aerocar Factory at Opa-locka

The Curtiss Aerocar factory at 1170 Sharazad, Opa-locka (ca. 1926 prior to occupation by Aerocar in 1928)

Interior image of the Aerocar factory at Opa-locka. This was far from a production line but still claimed an output of three Aerocars per week (ca. 1928).

An Aerocar nearing completion (*left*) with the frames of two Aerocars (*right*). Seating awaits installation on the right (ca. 1928).

Hugh Robinson (*seated, right*) at a workbench in the Aerocar factory (ca. 1928)

should also be acceptable to the patent office. Automakers willing to develop or adapt vehicles to tow the Aerocar would also be needed. A stock-model Hudson coupe was used for the first promotional trips. The Hudson would remain Curtiss's preferred tow vehicle, but others would soon follow. Once it had completed its promotional duties, Aerocar Number 2 would be repainted and delivered to Fisher as his first Aerocar.

The driver for the second trip was G. Carl Adams, with Fields as the relief driver. They had one passenger, Dr. A. W. Ziebold, the executive secretary of the Miami Chamber of Commerce. The Aerocar was an important new product from the Miami area and would be used to promote Miami to northern investors. To help him in this task, Ziebold carried with him a box of avocados for the mayor of New York and some roses for his wife.

Fisher's promotional skills were visible in all aspects of the trip. The Aerocar was provided with a police escort out of Miami, and its departure time was carefully selected to coincide with the departure of the Atlantic Coast Line train, perceived as a future transport rival of the Aerocar in long-distance travel. The Aerocar was twenty-two minutes ahead of the train at West Palm Beach, having maintained an average speed of 50 mph (80 km/h).

The journey to New York was completed on August 8 in thirty-nine hours and seven minutes. Goodyear Tires had an advertisement ready to publish on the day, confirming the journey time and of course the use of Goodyear All-Weather Tread Tires on the Aerocar.

During the remainder of 1928, several more promotional trips were undertaken by Aerocars. Automobile manufacturers Stutz of Indianapolis and Graham-Paige of Detroit were both keen to provide towing vehicles. In each case, a coupe had to be used to allow the fifth-wheel hitch of the Aerocar to be mounted at the rear without hindrance. Improbable claims made by Curtiss (later modified) that speeds achieved with the Aerocar in tow were higher than driving the automobile alone were no more than a gimmick but certainly didn't harm sales.

In November 1928, it was announced that the Opa-locka plant had been operating at its capacity of three Aerocars per week since inception and that products used in the manufacture of Florida Aerocars would, as far as possible, be local. In the year before the stock market crash of 1929, the Aerocar had had the best possible start.

THE FIRST PRIVATE AEROCARS

Such a trailer will answer the purpose of two Rolls Royces.

—Letter from Carl Fisher to Roy D. Chapin, July 12, 1928

The first Aerocars made at Opa-Locka in 1928 were for Aerocar shareholders, including Carl Fisher, Roy Chapin, and Howard Coffin. Fisher told all of his wealthy friends to buy one, and some did. Every early Aerocar was custom built, with each owner wanting different features or options. Fisher provided frank feedback to the factory on his Aerocars, berating the company for not making it easy for owners to uncouple the Aerocar and not including a third wheel for the Aerocar to be moved when not hitched to a vehicle. Very soon, however, the factory had ironed out the early bugs and was producing streamlined Aerocars of high quality. As orders increased during 1929, additional Aerocar design development was carried out in Hammondsport to allow Opa-locka to focus on production. Examples of early private Aerocars follow.

Curtiss with an early Aerocar and 1928 Hudson tow vehicle at Hammondsport. Early Aerocars had no wheel arches. Much design work was carried out at Hammondsport for early Opa-locka models (ca. 1928).

An Aerocar with 1929 Hudson (1929). The Aerocar was built at Opa-locka for H. H. Raymond, chairman of the board, AGWI (Clyde-Mallory) Steamship Lines, and shows considerable improvement in styling details over a short period.

A prototype Aerocar at Opa-locka with a swept rear end and forward roof extended over the driver's area (ca. 1929). An early design intent was to merge tow vehicle and trailer visually, creating the appearance of a single, triple-axle vehicle.

An Aerocar with a rear outdoor observation deck and canopy (ca. 1929). The tow vehicle is a Ford Model A.

The streamlined possibilities of the Aerocar were illustrated in a deluxe model with a rear end resembling a dirigible (ca. 1929). This Aerocar is being handed over to J. A. Seitz of Syracuse, New York (*middle*), with Curtiss engineer Hugh Robinson on the right. The tow vehicle is a 1929 Hudson.

CANDY, VEGETABLES, HORSES, AND GARBAGE—THE FIRST BUSINESS AEROCARS

In stark contrast to the highly visible, Fisher-inspired promotional tours of early Aerocars, Curtiss continued to use the Aerocar for its original intended purpose—to carry fruit and vegetables. The Aerocar was used for business before it was considered for leisure. Farmers and vegetable growers of the Opa-locka district acquired (or more likely were given by Curtiss) an Aerocar to deliver fresh produce three times a week to the leading hotels of Miami Beach. In February 1929, the Curtiss-Bright-owned Brighton Valley Farms advertised a free two-day scenic tour of the Everglades in an Aerocar. The purpose of the tour was to give a demonstration of the "vast agricultural possibilities of South Central Florida."

The first commercial Aerocar manufactured at Opa-locka for a third party was completed in October 1928 for traveling candy jobber Paul Ogden of Hialeah. It was not dissimilar in size and form to Aerocar Number 1. In March 1929, an Aerocar was built to transport four polo ponies belonging to two Foster brothers and John Walter who were wintering in Miami. The Fosters also purchased a touring Aerocar for their own use. In April 1929, the City of Miami assessed a simplified Aerocar for use as a possible garbage collection vehicle.

These low-key uses of the Aerocar were Curtiss driven and were largely for local community benefit. They did, however, demonstrate that the Aerocar could be used for a range of purposes, including light freight work, passenger and horse transport, and even garbage collection.

They also showed the importance that Curtiss placed on the developing Florida's back lots in support of city center and resort growth. Curtiss encouraged the use of his new transport invention to provide employment and food to the local community. If it also gave rides to polo ponies, so be it.

Examples of early business Aerocars are shown in the following pages.

Early Aerocar interiors

Early Aerocar interiors were basic, using wickerwork chairs, cushions, and exposed roof struts (ca. 1928).

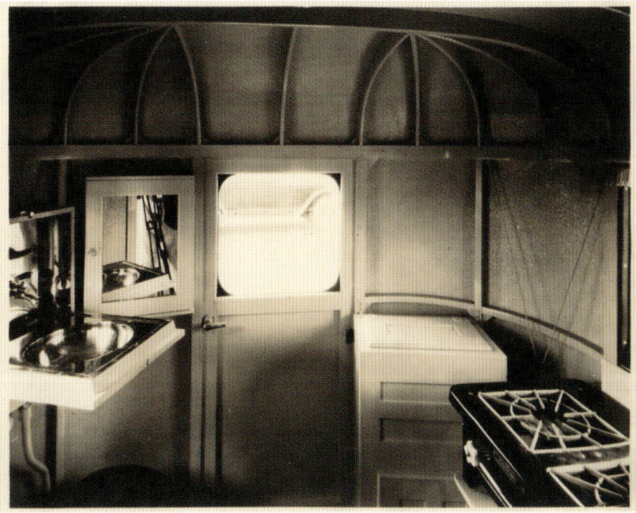

A kitchen area was available at the rear of some early Aerocars and included a folding sink, a two-burner stove, and an icebox (ca. 1928). A toilet was also available, although some customers complained it was placed too close to the kitchen.

Early ground transportation Aerocars included forward-facing wicker seats and viewing windows along both sides (ca. 1929). Linoleum was used for the floors, and luggage netting was installed in the roof.

This ground transportation Aerocar included newspaper racks and a radio at the front for the entertainment of passengers (ca. 1929).

The first business Aerocar completed for local candy jobber Paul Ogden (1928)

A polo-horse-carrying Aerocar built for the Foster brothers and John Walter (1929)

An Aerocar used for furniture transportation (date unknown)

Two interior photos of the Ogden candy car (1928)

DETROIT BEGINS PRODUCTION

By early 1929, more than twenty men were working at the Opa-locka plant, producing its full capacity of three Aerocars per week. It was time to expand.

Fisher's promotional tours in the North, combined with his contacts among the gasoline aristocracy of Detroit, created enthusiastic support for the Aerocar in the automobile-manufacturing industry and the press. Although Fisher's initial idea was to establish a number of manufacturers nationwide who would build Aerocars under license, only Briggs Manufacturing of Detroit was selected to build cars alongside the Opa-locka plant. Briggs produced bodies for Ford and other automobile manufacturers and was well regarded in the industry. Company records indicate that Briggs was the first to be contacted by Fisher, giving them a head start over others. As Briggs management realized the sales potential of the new trailer, they grew increasingly reluctant to let others share in the spoils. Manufacturing interest from other companies, including the Lang Body Company, the Mengel Body Company,

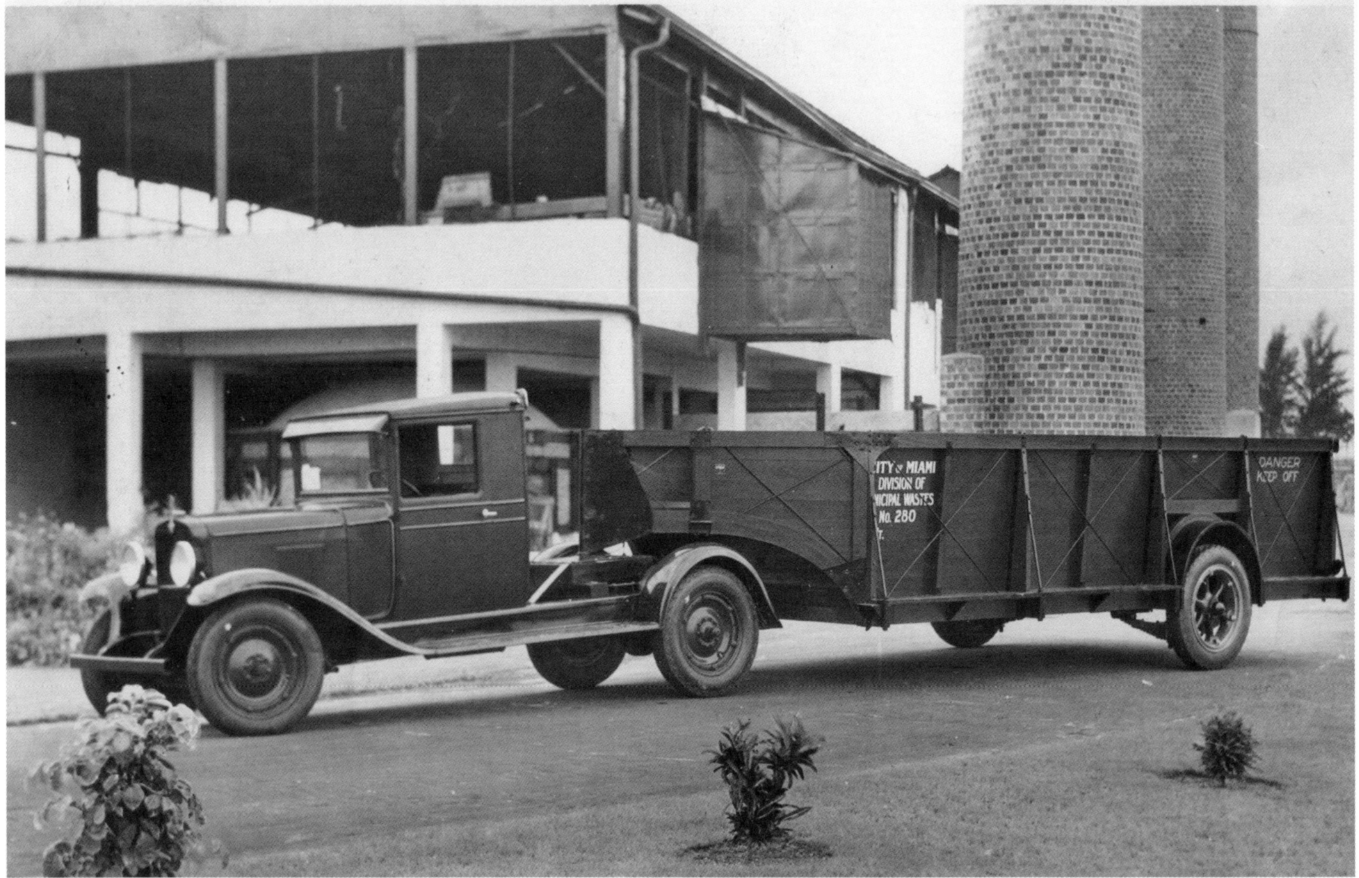

The garbage collection Aerocar and Chevrolet truck trialed by the City of Miami (1929). In some newspapers, it was called the "Aerotruck."

and the Weymann-American Body Company, was never converted into further license agreements. Company records suggest the existence of a license agreement with former racing driver and Fisher acquaintance Tommy Milton in Los Angeles, but no Aerocars are known to have been built in California.

The Aerocar Company of Detroit was formed in March 1929, with Briggs owner Walter O. Briggs appointing Byron F. (Barney) Everitt as president and general manager. A small factory was leased to produce the Aerocars at 7425 Melville Street, Detroit.

Detroit Aerocars were manufactured from broad designs supplied by Curtiss, modified by the Aerocar Company of Detroit. All design changes were subject to Curtiss's approval. Briggs designed a small family of Aerocars better suited to mass production at lower price points than the specialist vehicles made at Opa-locka. Designs were simplified, and there were few options. Over time, this distinction became clearer, since the Aerocars manufactured in Florida were the only ones to be given the name "Curtiss Aerocars." By August 1929, production was in full swing at the Detroit plant. Although the plant had a claimed capacity of thirty Aerocars per day, this number was never reached. Four different models were initially built in Detroit. They were a Standard Commercial Car ($1,000), a Standard School Bus ($1,200), a Standard Passenger Bus ($1,500), and a Standard Tourist Car ($1,500). Prices of custom models built in Opa-locka were far higher. Detailed prices and specifications of Miami and Detroit models are available in appendix C.

THE CUBAN CONNECTION

If this letter reaches you in time[,] won't you make an inquiry about the possibilities for the Aerocar in European countries. The boys just got back from Havana[,] where they say that it creates more interest and attention than a new 'plane. They think they will sell a hundred there this year.

—Letter from Curtiss to C. M. Keys,
August 2, 1929

Cuba was the first country outside America to see an Aerocar, although they were never sold in the numbers forecast by Curtiss. The second of these (the first was sold to the president of a Cuban brewing company in August 1929) was a gift to President Machado of Cuba from Curtiss, Carl Fisher, and Howard Coffin, all of whom had business interests in the country. Although the president's Aerocar appeared new, company records indicate that Aerocar #13, belonging to Carl Fisher, was extensively refurbished for this purpose, becoming Aerocar #113 in the process. It was delivered to the president in December 1929 as part of a broader sales push by Floridian businesses to promote tourism and commercial flights between Florida, Cuba, and South America. Other Aerocars later sold in Cuba were for the Sevilla Biltmore Hotel, Crusellas soap distributors, and Pan American Airways.

Curtiss's Cuban sales are an example of how the Aerocar quickly became a brand ambassador for Florida in general and Miami in particular. Florida soon became intensely proud of its Aerocars.

The Aerocar donated to President Machado of Cuba (1929)

THE AIRLINE CONNECTION

> Can't we start a campaign to get this job in use at all airports to meet passenger ships etc.? The easy riding qualities, space for baggage and mail, economy and the aeroplane construction should be arguments enough.
>
> —Letter from Curtiss to C. M. Keys, June 28, 1928

Investment banker C. M. Keys jumped at Curtiss's suggestion to provide ground transportation for his new airlines. Throughout the 1920s, Keys had remained at the helm of Curtiss Aeroplane and Manufacturing Company and had overseen the merger with the Wright brothers' enterprise to form the Curtiss-Wright Corporation in 1929. The "Lindbergh boom" that followed Charles Lindbergh's nonstop flight across the Atlantic on May 20–21, 1927, unleashed a new wave of interest in flying. Not only did more people want to fly on a plane, but bankers were now willing to invest in flying as a commercial proposition.

The Aerocar's first contribution to the emerging passenger airline industry of the late 1920s took the form of ground transportation for two new American airlines. During the 1920s, aircraft range improved and navigation made major advances, but long-distance commercial flying was not yet feasible, and nighttime flying was still risky. So airline passengers would need to be accommodated in hotels overnight.

Ground transportation was a key element in the success of early commercial air travel. To help airlines in competing against the railroads, their passengers would need to be whisked between hotels, railroad stations, and airports quickly and smoothly, almost as if they were still flying.

The first two airlines to use Aerocars as part of their ground transportation fleet, doubtless encouraged by Keys, were Transcontinental Air Transport (TAT) and Southwest Air Fast Express (SAFE). TAT investors included both Keys and Lindbergh and came to be known as "the Lindbergh Line." It purchased a fleet of eleven Aerocars for use along its coast-to-coast flying route between Los Angeles and New York. The TAT Aerocars were powered by Studebaker President cabriolets. SAFE purchased five Aerocars for use in the Midwest and Southwest, powered by Greater Hudson coupes. Pan American Airways would later purchase four Aerocars to transport passengers from business districts to airports.

Aerocar airline transport cars were luxuriously furnished with wicker chairs and seated up to thirteen passengers who could communicate with the driver by telephone. They were comfortable and quiet, far quieter than the noisy Ford Trimotor planes used on some routes in which cotton wool was handed out to passengers to protect their hearing.

TWO CRASHES

In 1929, press rumors continued to circulate about the possible manufacture of Aerocars beyond Miami and Detroit. In December 1929, it was stated in the press that the Curtiss Aerocar would be made in Phoenix, Arizona, by the Southwestern Aerocar Company, as well as a new product, the Aerotruck, which "is successfully adapted for garbage collection, cotton hauling[,] and various commercial purposes." The plant would start making two cars per week, with Hugh Robinson of Aerocar planning to move from Florida to Phoenix with his family to become chief engineer and general superintendent of the plant.

Along with earlier rumors of another new Aerocar plant to be built at Rochester, New York, these stories turned out to be wishful thinking. This was in part due to the Detroit plant not wishing to share Aerocar licensing with others outside Florida, but also the result of two crashes, neither of which involved Aerocars. The first was the crash of the TAT plane City of San Francisco on September 4, 1929, in a thunderstorm, killing all eight people on board. Tragically, the risks of flying had become plain for all to see, including potential passengers who deserted the new airlines for a time. The second was the Wall Street crash of October 1929, which heralded the start of the Great Depression. All speculative investments, which included Aerocar, ended, and the company's grand expansion plans were never fulfilled. It was a dramatic end to a year that had promised so much.

FROM POLO PONIES TO SHOES

In a twist of fate, the temporary collapse of passenger airline demand and the Wall Street crash had some silver linings for the Aerocar. Due to safety concerns, wealthy travelers reverted from planes to roads and railways for a time, and due to the stock market crash, salesmen had to work harder to get sales. Instead of customers coming into a store to buy products, the products had to travel to customers, especially in regional areas. The day of the traveling showroom had arrived.

The Enna Jettick Shoe Company of Auburn, New York, realized that in the postcrash economy, a traveling shoe showroom made good business sense. They

An Aerocar and 1929 Hudson forming part of a fleet owned by Pan American Airways with a Ford Trimotor plane in the background (ca. 1930)

decided that the Aerocar was the best mobile showroom on the market, and purchased at least four of them beginning in December 1929.

The rear end of the Enna Jettick Aerocars was designed to look like an airship, a design also used on some private Aerocars. The partnership between Aerocar and a shoe company was a win-win. Enna Jettick would stand out from a crowd of shoe manufacturers, while Aerocar would have a traveling showroom of its own, always on the road and always seen by thousands. It was a business model that Aerocar would employ successfully with many other companies throughout the 1930s.

Another shoemaker, the Jarman Shoe Company of Nashville, used an Aerocar called the "Friendly Five Flyer" to market its Friendly Five Shoes for Men. Their shoes were "friendly" because none were priced over five dollars.

A Friendly Five Shoes Aerocar generating much interest (ca. 1930)

EARLY AEROCAR ADVERTISING AND LOGOS

Throughout its life, the Aerocar would rely on the free publicity generated by its diverse and loyal owners. But Curtiss, and most likely Fisher too, realized at an early stage that the company needed to keep control of its brand and messaging through selective advertising.

Unlike the Opa-locka plant, which encouraged buyers to buy directly from the factory, the Detroit factory used automotive dealers to sell Aerocars.

The interior of an Aerocar mobile shoe store (ca. 1930)

Dealers knew that buyers interested in the Aerocar might buy a new tow vehicle too, and it was they who first advertised the Aerocar. Adverts began appearing in northern newspapers in October 1929, around the time of the Wall Street crash. Business uses of the Aerocar featured strongly in Detroit advertising.

When the Florida plant began to trial its own advertising in 1930, it sent an altogether more sedate message, emphasizing the aeronautical heritage of the Aerocar and the custom-built nature of its product. Examples of Aerocar advertising are in appendix B.

Aerocar logos were not particularly elegant or eye-catching, but they would always remind customers of the Aerocar's aviation heritage. Aerocar in Florida would use a stylized bird's wing alongside an automobile tire, while Detroit hedged its bets by using both a set of bird wings and an arrow. The Florida logo was updated in 1933 to incorporate the theme of "modern transportation." The disparity in branding between Florida and Detroit highlighted ongoing differences between the two companies and would not have helped sales in a market that by the mid-1930s was being flooded with travel trailers of all kinds.

1930: THE DARKEST YEAR

As the Great Depression took hold and workers lost first their jobs and then their houses, many thousands moved permanently into travel trailers, taking to the road to look for work or simply as a cheap place to live. This was to become one of the greatest social upheavals in twentieth-century America. It seemed as if half the country was on the move. But due to cost, few would do so in an Aerocar.

The first Aerocar logo showed a stylized wheel and wing-shaped trailer (ca. 1929).

Curtiss's name was later added to the wheel and wing logo (ca. 1930).

The Aerocar's logo was changed about the time of Aerocar Florida's move from Opa-locka to Coral Gables. It introduced the concept of "modern transportation" to reflect the wide range of uses of the Aerocar along with a "spread eagle" logo. This design was also used on Aerocar badges (ca. 1933).

The wheel, wings, and arrow logo used by the Aerocar Company of Detroit (ca. 1929)

Still, the first half of 1930 began well for the company despite the major shock of the Wall Street crash. Aerocar traveling showrooms were seen by many businesses as a smart response to the retail recession. New uses were being found for the vehicles around the country, including an ambulance, a "church on wheels," and an Aerocar used to campaign in Pennsylvania against Prohibition. The Aerocar was surviving the economic storm, and prospects looked good.

Then, out of the blue, on July 23, 1930, Glenn H. Curtiss died in New York of a pulmonary embolism while recovering from an appendicitis operation.

An Aerocar belonging to the Peoria Casket Company (ca. 1930)

CHAPTER 6

THE CURTISS AEROCAR (1930–40)

[Before the Aerocar], no one had brought to the nondescript assortment of motorless trailer vehicles the vast store of engineering improvements adaptable to the semi-trailer. It had remained a homemade product, suffering all the faults of makeshift arrangements, incompetent engineering, and not infrequently a menace to traffic. It had not kept pace with public demands in the matter of comfort, convenience, speed and economy.

—Aerocar publicity (1935)

Opposite: The Aerocar and International tractor in the hands of famed industrial designer Brooks Stevens. This Land Cruiser was designed by Stevens for the Plankinton Company (built 1936, photographed 1938). *Courtesy of Brooks Stevens (American, 1911–95), Plankinton Company, and Land Cruiser, 1938; gift of the Brooks Stevens family, the Milwaukee Institute of Art and Design, and Milwaukee Art Museum Institutional Archives, BSA_0037*

THE END OF THE CURTISS ERA

Curtiss's death came as a complete shock to nearly everyone who knew him. Financier C. M. Keys had foreshadowed some of the problems Curtiss might face in Florida when he wrote to Curtiss in 1926:

> You are most fortunate in that your properties have, on the whole, been well bought. I think there are too many of them and that you would be very much happier if you had only Opa-Locka or Country Club Estates on your hands, and I hope you will not get into any more big adventures until these begin to pay out in substantial amounts, because I know that in spite of your complete confidence in your enterprises and in the personnel running them, you must nevertheless feel the strain of these great business enterprises at times quite severely.
>
> —Letter from C. M. Keys to Curtiss, November 29, 1926

Three years later in 1929, Keys wrote in a more serious tone:

> Since my first contact with your Florida problems I have been much concerned by the accumulated evidence of poor judgement and mismanagement evidenced in the Hialeah Bank, and in Brighton, and I shall be appreciative of all information which can be given me with regard to these two projects.
>
> —Letter from C. M. Keys to Curtiss, September 25, 1929

The second letter was written a month before the stock market crash. From this and other correspondence, it is apparent that despite a bright start in Florida and Curtiss's generally conservative real estate investments, he had over time been negatively affected by regular flooding at Hialeah, the end of the Florida real estate boom (1925), two hurricanes (1926 and 1928), and the stock market crash (1929). Curtiss was overextended, probably overgenerous, and overworked. In 1930 there was no credit available for any Florida-based business regardless of its pedigree. Severe headwinds had turned into a brick wall.

The Aerocar was a small glimmer of light in a dark tunnel, but the business was neither mature nor substantial enough to compensate for Curtiss's real estate losses. Compounded by the stress of an ongoing trial with former business partner Augustus M. Herring dating back to 1910, these events must have affected Curtiss's physical and mental well-being deeply in 1930.

A LEADERLESS TEAM

On July 25, 1930, business was suspended at Curtiss plants in Hammondsport and Opa-locka for Curtiss's funeral. Dignitaries from around the country came to Hammondsport to pay their respects. As the world paid homage to a great man, there was little space in Curtiss obituaries for the quirky vehicle he had invented toward the end of his life. But Charles L. Lawrence, vice president of the Curtiss-Wright Corporation in charge of engineering and research, found some room:

> His recent work in the development of the Aerocar, a motor car modelled on the lines of the airplane, has unfortunately been cut off at a point where it was just about to be perfected.
>
> —*Times Union*, July 27, 1930

For a time after Curtiss's death, the Aerocar team at Opa-locka lost motivation. On August 15, 1930, Fisher wrote to Keys, "Since we have lost Mr. Curtiss, things generally seem to be at a standstill." On January 27, 1931, Keys lamented the lack of profits at the Opa-locka plant. He wrote to Fisher, "I do not think the Florida company will ever produce anything except good cars and wonderful deficits." The company's owners were similarly preoccupied. Fisher wrote to Howard Coffin on December 10, 1931, that "all of our stockholders[,] in fact, are too wealthy in their own names, and too busy in their own business, to pay any attention to the Aerocar patents or industry."

The Great Depression combined with Curtiss's death the following year exposed rifts within the complex and geographically dispersed Aerocar organization. There were three disparate groups: the "Curtiss interests" in Florida, comprising loyal Curtiss engineers, family members, and friends who saw the Florida plant as an "experimental station" and who wanted to produce and refine high-quality Aerocars in small quantities regardless of cost; the "New York interests," consisting of lawyers and financiers who managed the licensing company and who wanted greater licensing income and more robust patent enforcement; and the Detroit company, which was producing Aerocars independently for a fixed price that were by Fisher's account of poor quality and not selling well.

The Florida company was being pulled in different directions. Fisher wanted it to explore the use of small railcars fitted with Aero Coupler–style connections, but after the production of a single prototype, this was not pursued. He complained to H. Sayre Wheeler, the new head of Aerocar in Florida, about the constant changes in Aerocar design (which Curtiss would have no doubt encouraged). On April 4, 1931, he wrote to Wheeler, "I don't see how you can expect me to be the experimental goat on your inventions."

A more affordable "Camper's Express" Aerocar introduced by the Aerocar Company of Detroit in 1931 to respond to the Depression (1933). The tow vehicle is a 1930 Model A Ford. *Courtesy of the Gleason Waite Romer Photographs Collection, from the Miami-Dade Public Library System*

By 1931, it was clear that the Aerocar was facing extinction unless some new blood could be injected into the company. The Aerocar was a luxury discretionary item looking for customers in the midst of the worst economic depression the country had seen.

NEW FACES

In March 1931, the Aerocar Company of Detroit was sold to John A. Schroeder and W. J. Parrish, two Detroit automobile executives who saw the potential of the Aerocar as a lower-cost camping trailer as well as a mobile showroom for products and services. The company's new owners agreed to buy the company on condition that no further Aerocar manufacturing licenses would be issued. Although Fisher later described this deal as a "huge mistake," the preoccupied, recession-hit Aerocar owners agreed—beggars could not be choosers. It's not clear if the Detroit company's previous owners were forced out due to quality concerns or if this was simply one of many corporate reorganizations in response to the Great Depression. Whatever the reason, Briggs continued to manufacture the vehicles in Detroit under new ownership. The new license agreement stipulated the manufacture of Aerocars for three years, with reduced royalties on a sliding scale starting at $40 per vehicle subject to a minimum of 250 cars in the first year, 350 in the second year, and 500 in the third. It is highly unlikely these targets were met.

That same year, financier C. M. Keys took a back seat in managing the company from New York. He would continue to support Aerocar informally but was heavily involved in managing his interests in the passenger airline industry, which was badly affected by the Wall Street crash. In 1932, he reduced his workload due to ill health to concentrate on his property interests.

In Florida, H. Sayre Wheeler remained as president, with Curtiss engineer Hugh Robinson as vice president in charge of production. G. Carl Adams shifted his focus to local government matters, became mayor of Miami Springs in 1930, and would no longer have a prominent role in the manufacturing or sale of the Aerocar. Unable or unwilling to invest further in Aerocar, the New York shareholders began looking for someone to take the rest of the Aerocar organization off their hands.

THE ROBINSONS

Hugh A. Robinson and his son Harold H. Robinson were both closely associated with the Curtiss Aerocar Company in Florida and played important roles in the company after Curtiss's death. With the exception of the patents filed posthumously on his behalf by Curtiss's widow, Lena, almost all of the patents issued by the company during the 1930s were in the name of Harold H. Robinson.

Hugh Robinson was a longtime friend of Curtiss and a fellow pilot. After being lent a propeller by Curtiss that enabled him to fly a plane successfully at the St. Louis Centennial Exposition in 1909, Robinson joined Curtiss in 1911 as engineer, chief pilot, and member of the Curtiss exhibition flying team. He went on to spend time in France selling Curtiss hydroplanes and later left Curtiss to work for the Aero-marine Airplane Company. Robinson developed the "hook and cable" arresting-gear system, used to brake planes landing on aircraft carriers.

In 1924, Robinson was looking for employment and contacted his old friend Curtiss. As he did with a number of other former employees, Curtiss offered him a job helping to develop Opa-locka in Florida, where Robinson became a member of the local council and the town's fire chief. When Aerocar construction began in Florida in 1928, Hugh Robinson was the natural choice to oversee the production side of the business. He became the vice president, chief engineer, and supervisor of the Opa-locka plant. It is likely that Robinson would have been involved in the engineering and construction of the Brighton trailers, the first Aerocar, and the Aero Coupler. Robinson Sr. continued to supervise Aerocar construction in Florida until its demise in 1940.

Hugh Robinson's son Harold (Hugh had a second son, Hugh Armstrong Robinson Jr.) became president of the Curtiss Aerocar Company of Florida in 1932, shortly before its move from Opa-locka to Coral Gables in 1933. In March 1937, he was called in the press the "guiding brass hat of the Rolls-Royce of trailerdom." Harold had a keen interest in amateur radio and was an innovative engineer in his own right.

THE AEROCOACH (1930)

The Aerocoach was the first new Aerocar product to be launched after Curtiss's death. Designed by Curtiss, it was left to Hugh Robinson to turn Curtiss's concept into reality. Using a Ford truck chassis, it was a more sophisticated commercial application of the Trailer Bus of 1922. The first model had a capacity of twenty-one seated and fifteen standing passengers. As with the Trailer Bus, the Aerocoach driver had a separate compartment but was still able to collect fares and operate the doors. The Aerocoach weighed 4,000 pounds and had a top speed of 60 mph (97 km/h) and a claimed operating cost of ten cents a mile.

Hugh Robinson at the controls of a Curtiss Model D plane (1914)

External and internal views of the Aerocoach (1930). Its main party trick was to give the driver of the articulated bus access to the passenger compartment for the collection of fares.

Although the Aerocoach is known to have been trialed by the Miami Traction Co. in northwest Miami in late 1930 and by General Motors in 1931 for use as a Greyhound bus, there are no records of the Aerocoach going into extensive public use. It is likely that the lightweight construction methods used for the Aerocoach offered inadequate protection in the event of a side impact or rollover. Passenger safety was a matter of growing importance from the early 1930s as road traffic and accidents increased.

A promotional photo for the Aerocar transportation service established by Henry Doherty to shuttle guests between his Miami-based hotels and clubs (1933). *Courtesy of the Gleason Waite Romer Photographs Collection, from the Miami-Dade Public Library System*

HENRY L. DOHERTY AND THE YEAR-ROUND CLUBS

A man wants a bird's-eye view of things as he travels and all he gets out of automobiles is a snake's belly view.

—Henry L. Doherty as reported in his obituary in the *Battle Creek Enquirer*, December 31, 1939

In 1932, after talking to at least two other interested parties, Aerocar's initial stakeholders found their exit strategy from the company. Both the Aerocar holding company and its Florida operations were sold to utilities magnate and Florida hotel owner Henry L. Doherty.

From the late 1920s, Henry L. Doherty (1870–1939) was a major investor in Miami tourism infrastructure and became an ambassador for investment in the Florida region. Often called an oil man, he had extensive interests in oil, gas, and electricity. He owned Cities Service Company, an oil services company worth over $1 billion in 1929. By 1936, he was a director of ninety-five corporations and president of eight-seven of them. Doherty suffered from arthritis throughout his life. Based in New York, he visited Miami often to take advantage of its warmer climate.

Doherty helped Miami recover from the 1925 real estate collapse and 1926 hurricane by investing heavily in hotels, sports clubs, and other tourism ventures. He purchased the Roney Plaza and Miami-Biltmore hotels in 1929 and was president of the Florida Year-Round Clubs, a group of local sports clubs and organizations who came together to promote year-round visitation of Miami. To facilitate transport links between his properties, he decided to use Aerocars. Doherty was sufficiently wealthy not just to buy a fleet of Aerocars but also to buy the company.

After Curtiss's death in 1930, the Aerocar Company of Florida became part of the Curtiss estate, slowing down the legal process of asset disposals. It would take two years to unravel Curtiss's business affairs. In mid-1932, Doherty acquired the Aerocar Company of Florida, bought two Aerocars for himself, and, in October 1932, purchased a fleet of twelve Aerocars for $75,000 to provide free ground transportation between his Florida tourist properties, clubs, stations, and beaches. The Doherty fleet of Aerocars would later increase to sixteen. He helped promote the Aerocar brand by lending his private Aerocars to celebrities including golfer Gene Sarazen for long-distance travel. Doherty's Aerocar fleet was described by the *Miami News* as "the most palatial system of motor transportation ever offered in the country." The Miami-Biltmore hotel also used a Pitcairn PA-19 autogiro as part of its customer fleet. Doherty's Cities Service Company would also use business Aerocars and a gasoline-carrying truck called the Aerotank.

THE DOHERTY AEROCAR (1932)

In October 1932, an Aerocar was delivered to Henry L. Doherty that represented the pinnacle of the company's achievements to that date. It was built at Opa-locka for Doherty's personal use and was made at the same time as the rest of the Doherty Aerocar fleet. It was far more lavish than the ground transportation models used by airlines, and was the first Aerocar to include an upper-level observation deck. A description appeared in the *Miami Herald* in October 1932 that is worth quoting in full:

> It is the most expensive and modernly equipped to be built by the company. The main passenger compartment, upper left, provides seats for four persons, upper and lower seating berths, with two additional in the rear observation compartment, which also includes a complete concealed kitchen, with ice box, gas plate, porcelain sink, pressure water system, hot wat heaters, folding table[,] and disappearing shower bath. Other equipment includes electric fans, built-in radio, electric clocks, loud speakers. Upper right is the airplane[-]type observation compartment with instrument board on which are installed barometer, altimeter, magnetic compass, searchlight controls, microphones, thermometer[,] and other units. It is enclosed by plate glass. It is powered by a Graham eight coupe. Exterior equipment includes window ventilators, ultra-violet ray glass in roof, luggage carriers, electric brakes, all-round bumper[,] and Firestone gold standard heavy duty casings.
>
> —*Miami Herald,* October 23, 1932

Curtiss Aerocar Company, Inc.

We Compliment You

On the Design, Construction and Completion of America's Finest Piece of Motor Travel Equipment—

The Private Aerocar of Mr. Henry L. Doherty

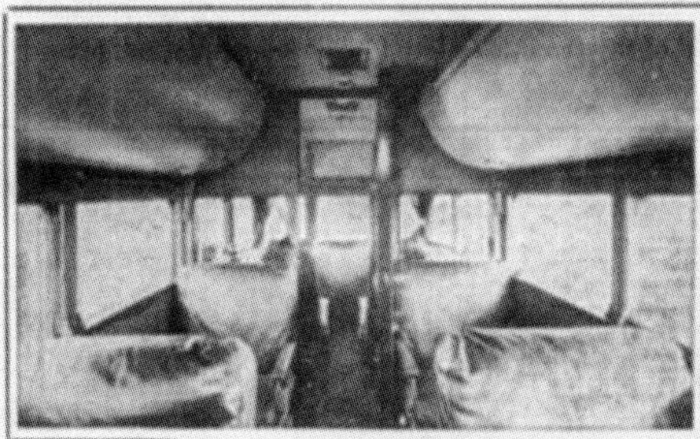

Built By Curtiss Aerocar Co., Inc. Opa-Locka, Fla.

Powered By Graham Great "8" Coupe

We Are Glad That Miami Produced This Aerocar and That We Co-operated In Its Production

"*Equipped*" *With*
FIRESTONE HEAVY DUTY GUM-DIPPED TIRES
FIRESTONE SERVICE STORES, INC.
1200 WEST FLAGLER STREET

DuPONT TOP MATERIALS Furnished By
AUTO LUGGAGE & TRIMMERS' SUPPLY
1211 N. E. SECOND AVENUE

RENUART LUMBER YARDS, INC.
"Everything To Build Anything"
CORAL GABLES, FLORIDA

EGYPTIAN LACQUER—SPRING WORK BY
MIAMI PARTS AND SPRING CO.
35 TO 41 N. W. FOURTH STREET

LIBBY-OWENS-FORD SAFETY GLASS BY
BINSWANGER & COMPANY
1212 N. E. SECOND AVENUE

POWERED BY GRAHAM "8" COUPE
STRATTON MOTORS, INC.
Distributors
301-315 SOUTH MIAMI AVENUE

PAINTING ON AEROCAR AND POWER UNIT BY
NAT T. McGEHEE, AUTOMOBILE PAINTING
"The Better Kind"—Seven Years In Miami
285 N. W. 80th ST. TEL. EDGEWATER 1693-M

THE SHERWIN-WILLIAMS CO.
Paints—Varnishes—Lacquers—Enamels
56 WEST FLAGLER STREET

We Are Delighted That This and All Other Curtiss Aerocars Are Equipped With
UNION BATTERIES
UNION BATTERY MFG. CO., 847 S. W. EIGHTH STREET

AS USUAL—CARPETS BY
MIAMI RUG CO.
22 SOUTH MIAMI AVENUE

PHILCO TRANSITONE AUTO RADIO SOLD BY **ELECTRIC SALES AND SERVICE**
IN ALL AEROCARS OF THE DOHERTY FLEET — 1121 N. E. SECOND AVE.

An advertisement placed in the *Miami Herald* on October 30, 1932, congratulating the Aerocar Company of Florida on the completion of the Aerocar for H. L. Doherty. *Courtesy of www.newspapers.com*

An advertisement placed a few days later by the companies who supplied components for the Doherty Aerocar gives an interesting overview of Aerocar suppliers at the time. They included Firestone, DuPont, Union batteries, Philco radios, Libby-Owens-Ford safety glass, and Sherwin-Williams paints

On the surface it appeared that in the midst of the Great Depression and postboom slump in Florida, Doherty had become Aerocar's financial savior. Aerocar employees in Florida had a different view. According to a letter written by Aerocar president and loyal Curtiss employee H. C. Genung in May 1937, during Doherty's period of ownership Aerocar had been prohibited by Doherty from selling any Aerocars to his competitor hotels. Furthermore, according to Genung, the company had made no money from Aerocar sales to Doherty, and Doherty had refused to invest any new capital in the company for ongoing working capital and research needs. In purchasing Aerocar Florida, Doherty was probably just protecting his hotel interests, seeing no reason to invest further in a highly specialized company that would have been no more than a footnote in his balance sheet.

But by 1932, the company had recovered from the death of its founder and under new ownership remained optimistic about the future.

The Aerotank as featured in National Petroleum News, October 11, 1933

THE AEROTANK (JULY 1933)

In 1933, Harold Robinson produced another Aerocar variant, this time for gasoline transport. He felt that the Aero Coupler suspension would lend itself well to the transportation of liquids, including sensitive materials such as gasoline, while the monocoque construction of the Aerocar could be used in commercial vehicles to reduce weight. Constructed of an Alcoa aluminum alloy, the Aerotank eliminated the need for a standard chassis, reducing the center of gravity and increasing safety.

The Aerotank was manufactured at the Coral Gables plant and was first used by Doherty company Cities Service Co. to transport a less volatile liquid, water, to the Key Largo Anglers' Club. Robinson applied for a patent for the tank trailer in July 1933, which was approved in April 1936.

Well-known airline passengers photographed in Miami with Pan American Airways Aerocars (1934). *Courtesy of the Gleason Waite Romer Photographs Collection, from the Miami-Dade Public Library System*

General Harold Fiske, commander general, US Army Canal Zone

Dean Radcliffe Heermance of Princeton University

Calixto Garcia Velez and his wife, Carmen Mendieta, the daughter of the Cuban president

George Harvey, British member of Parliament

RAVE REVIEWS

During the early 1930s, despite ownership uncertainty and ongoing tensions between the company's investors and staff, Aerocar's products were widely and positively reviewed across the nation. With Doherty's "club cars" a common sight in Miami from late 1932 onward, they became the transport symbol of the city. Local newspapers proudly reported on the Aerocar's national popularity and featured articles both on business and private owners of Aerocars. "Traction experts" called the Aerocar "revolutionary," and "housewives" commented on how easy it was to prepare meals inside the Aerocar on the move. "Never before has a motorless vehicle been elevated to such high standards" was a common refrain, and comparisons with Rolls-Royce automobiles were frequent. Special mention of the Aero Coupler was nearly always made.

Some reviews continued to peddle towing dynamic nonsense about the Aerocar. In February 1931, *Distribution and Warehousing* magazine claimed it was possible "to take curves at higher speeds than would be possible with the power plant alone" (such experiments should never be attempted with anything in tow). Despite its obvious aircraft roots, reviews and some of Aerocar's own publicity continued to call it a "land yacht" or a "houseboat of the highway," following the better-understood terms of the day. In any event, in the midst of the Great Depression, any publicity was good publicity.

MODERN TRANSPORTATION (1933)

By 1933, Doherty's acquisition of Aerocar had injected new enthusiasm and new orders into the company. Aerocar had recovered sufficiently to consider expansion. In March 1933, the Aerocar was exhibited at the Auto Show at the Coral Gables Coliseum and was the self-proclaimed "star of the show." In July, the Curtiss Aerocar Co., Inc., moved from Opa-locka to Coral Gables, a fashionable district of Miami established by George E. Merrick. The company explained the move in an advertisement: "The upward trend in national business and the increasing number of inquiries and orders have made it necessary to increase manufacturing facilities." The new premises covered 45,060 square feet (4,186 m^2) and included manufacturing and patent departments, woodworking, upholstery, power unit, shop, and stock and display rooms. Interestingly, the Coral Gables workshop appears to have retained the bespoke character of the Opa-locka plant, with no indication of mass production techniques. Clearly the ghost of Curtiss was still present in these new facilities.

From mid-1933 onward, Aerocar's national advertising took on a more sophisticated tone, with the slogan "Modern Transportation" used extensively. The Aerocar was described as a "drawing room on wheels" that "preserved its intimate and cordial mood" while up to a dozen guests were entertained with music, bridge, cocktails, and canapes.

The new Aerocar factory and showroom at 300 Valencia Avenue, Coral Gables, Florida (1933)

An external view of the Coral Cables factory

The Coral Gables showroom

The Coral Gables workshop

NEW MODELS

It was reported that one of Doherty's tangible contributions to Aerocar was standardization. Following his involvement, custom builds declined in number, and fewer models were produced, to reduce costs. Still, by 1935 the Aerocar was available in eight models. In 1936, Aerocar in Detroit introduced two lower-priced models, the $2,250 Curtiss Auto-Annex and the $1,425 Curtiss Gypsy. Although retaining the Aero Coupler, the Auto-Annex was marketed as suitable for use with either coupe or sedan, suggesting modifications to the Aero Coupler's design and position for sedan use. In 1937, a new model 161B was introduced, which included the observation deck developed for Henry Doherty in 1932.

The Graham Blue Streak Aerocar first owned by banker Hugh McDonald. *Courtesy of the Louwman Museum, The Hague, Netherlands*

The interior of the Graham Blue Streak Aerocar. *Courtesy of the Louwman Museum, The Hague, Netherlands*

PRIVATE AEROCARS

The list of individuals known to have owned an Aerocar in the 1930s is extensive. They were used for commuting, entertaining, and touring. It was not unusual for Aerocar owners to share their business commute with others. In 1937, four businessmen from Newark, New Jersey, shared an Aerocar on their regular commute to the city. According to one report, "They like it far better than commuting by automobile because they can't see ahead and are not under the constant strain of mentally applying the brakes."

Industrial designer Brooks Stevens added his magical touch to an Aerocar for the Plankinton Company in 1936. In a chapter titled "Battle of the Behemoths" from Myron Zobel's book *The 14-Karat Trailer* (1955), Zobel describes an impromptu road race between his Schult Continental Clipper and the Aerocar of William Woods Plankinton III. Traveling at up to 70 mph (113 km/h), the Aerocar won easily.

Some examples of private Aerocars manufactured in the 1930s follow. Additional examples can be found in Appendix D.

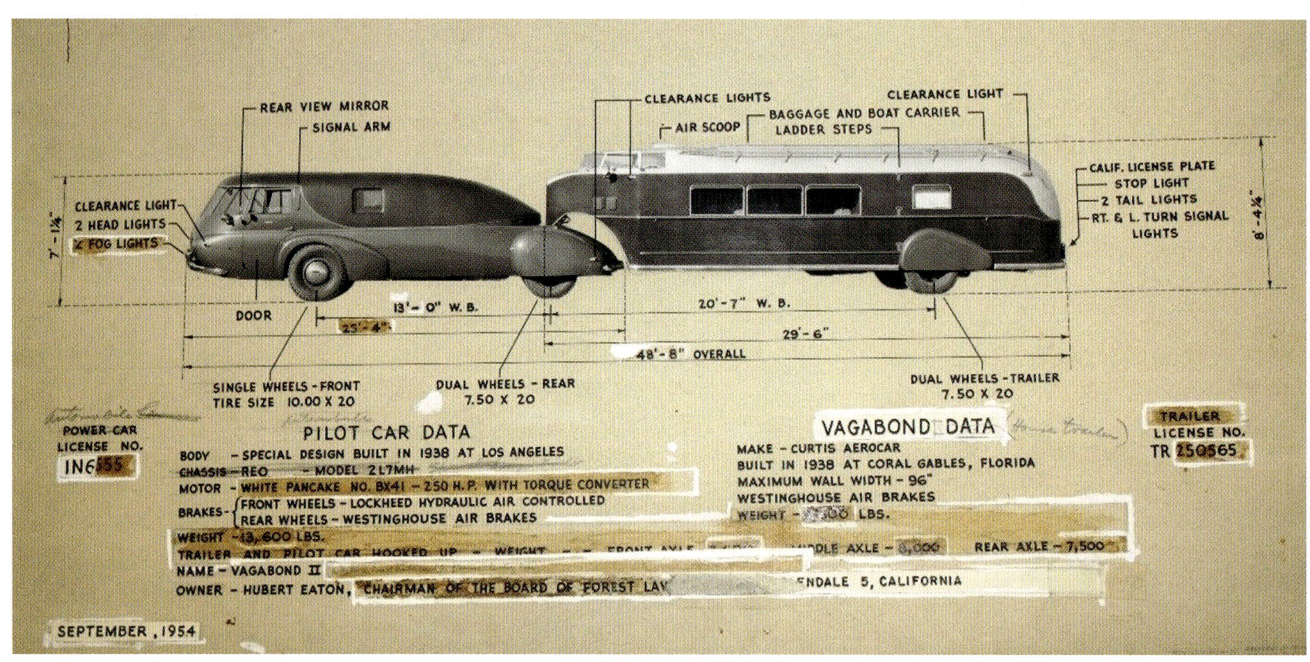

Museum-prepared specifications of the 1938 Reo and Aerocar. *Courtesy of the Petersen Automotive Museum, Los Angeles*

The 1938 Reo and Vagabond Aerocar of Dr. Hubert Eaton of Forest Lawn Memorial Parks, Los Angeles. The tractor was constructed by the Standard Carriage Works of Los Angeles. *Courtesy of the Petersen Automotive Museum, Los Angeles*

The 1938 Aerocar belonging to Mr. and Mrs. W. T. Samson Smith of Cooperstown, New York, and Short Hills, New Jersey, on exhibition at the Glenn H. Curtiss Museum, Hammondsport, New York

COMMERCIAL AEROCARS

From the outset, business sales were crucial to the success of the Aerocar. During the 1930s, trailer manufacturers would quickly learn that commercial sales could always be made even when the private sector was quiet. The Aerocar's flexible layout and ability to carry (for a trailer) significant payloads appealed to a wide range of businesses. Stretching the aircraft metaphor a little, Aerocar advertising in 1934 had invited salesmen to "ride through sales resistance with a Curtiss Aerocar." Aerocar's success in this sector was such that by 1938, Aerocar in Detroit claimed to be America's largest commercial trailer builder. This claim would have been disputed by Covered Wagon and Schult Trailers, to mention just two other large trailer makers of the period that also supplied the commercial market. With its in-depth supply of automobile parts manufacturers, Detroit was particularly suited to building commercial models to a fixed price.

The Norge Nestor Refrigerator Caravan (date unknown)

The General Electric Refrigerator Display Aerocar with a 1930 Hudson (1932). *Courtesy of the Gleason Waite Romer Photographs Collection, from the Miami-Dade Public Library System*

The Aerocar of the San Antonio Drug Company (date unknown)

The Aerocar of National Amplifying Systems (date unknown)

In addition to the airlines and club cars profiled earlier, some key purchasers of commercial Aerocars in the 1930s included the following:

Fostoria Glass Company of Moundsville, West Virginia, purchased six Aerocars in 1931 to showcase its products. The Fostoria Aerocars featured wooden cabinets and satin-lined drawers containing 1,587 glassware products that needed no additional packaging during transport.

Servel Inc. announced in February 1937 that it was putting one hundred Dodge coupes and Aerocar trailers into operation. The trailers were used as showrooms on wheels to promote Electrolux kerosene refrigerators to remote areas where neither electricity nor gas was available.

General Electric had by 1937 a fleet of over fifty Aerocars containing model electric kitchens. The fleet was used to promote electric appliances to public utility companies and consumers. Motive power for the GE Aerocars was supplied by Reo.

Cities Services Co. (owned by Henry Doherty) acquired two Aerocars in 1934 called "Cities Service Power Provers" to test gasoline engines and reduce carbon monoxide poisoning.

RCA Victor purchased two Aerocars in about 1936, powered by Hudson coupes, to demonstrate its products. They cost over $25,000 each and incorporated, according to RCA advertising, "the most powerful public address system put on wheels."

HEALTH CARS

Thanks to its shock-absorbing suspension, the Aerocar was ideally suited to carrying fragile objects and people, including medical equipment and hospital patients. Aerocars were converted into ambulances, mobile dental clinics, accessible buses, and mobile health education cars. In 1938, a special Aerocar was built in Coral Gables for juvenile paralysis victim Fred B. Snite Jr. of Chicago. Snite was permanently encased in an iron lung, seeing the outside world only through mirrors. His custom Aerocar included a revolving periscope to allow him to see in all directions. The Snite Aerocar was a familiar sight at football and baseball games and at horse races.

A public-health Aerocar clinic used to fight the war on syphilis in Georgia (1937). *Courtesy of Library of Congress (control numbers 2016872221 and inset 2016872222)*

OVERSEAS ORDERS

Beyond limited initial demand from Cuba, by June 1931 the Aerocar Company of Detroit reported demand for the Aerocar from "all over the world." This was a slight exaggeration. Examples given were the use of Standard Oil Company Aerocars in "Asia Minor" and Aerocar "buses" that were in the daily service of the Nairn Eastern Transport Company of Beirut and Damascus. Only two Aerocars are known to have been trialed by the Nairn company. Aerocar sales in Europe were handled by a representative of the Curtiss-Wright Corporation, but no sales were recorded there, more than likely due to the vehicle's size and cost. One British caravan magazine called American travel trailers "unhomely" because they lacked soft furnishings, carpets, curtains, and other trappings of a domestic environment. The Aerocar's main handicap overseas, however, was its fifth-wheel configuration, little known and even less understood outside America in a leisure context.

RUTH BRYAN OWEN

Ruth Bryan Owen (*left*) and daughter Helen Rudd Owen with their Aerocar (ca. 1930). *Courtesy of the Gleason Waite Romer Photographs Collection, from the Miami-Dade Public Library System*

Miami resident Ruth Bryan Owen was a film industry pioneer and later became the first woman elected to Congress from Florida, serving from 1929 to 1933. She was also the first woman to serve at the ministerial level in the US diplomatic service. In 1930, she took a 6,000-mile (9,656 km) family vacation in an Aerocar to the western states, including Yellowstone National Park. In 1931, she took an Aerocar to Denmark on a working vacation to learn about the Danish cooperative dairy-marketing system, with a view to applying it to Florida's fruit-growing industry. Owen later wrote a book titled *Denmark Caravan* about her travels and continued to use an Aerocar to visit towns in her Florida constituency. Owen was happy to be an Aerocar ambassador on her travels, and the brand would have received a boost from her association with the company. Company correspondence indicates that Owen's Aerocar was a Curtiss gift.

AEROCAR LOOK-ALIKES AND PATENT DISPUTES

I am beginning to think, however, that the best plan is to let the patents go, and go after the business.

—Glenn Curtiss, reportedly speaking about his patent dispute with the Wright brothers (1912)

Although imitation is often the sincerest form of flattery, for Aerocar it caused nothing but headaches. The Aerocar patents were not as valuable as first thought, and little time or money was dedicated to protecting them.

As the American travel trailer market expanded rapidly during the 1930s, the Aerocar was not immune from competitors seeking to replicate Aerocar looks at lower cost. The Aerocar patents applied for while Curtiss was alive focused on the Aero Coupler and Aerocar construction methods. When trailers that used different construction and connection methods but were similar in appearance to the Aerocar started appearing on the market in 1931, the company hurriedly applied retrospectively for additional design patents under the name of Lena Curtiss. But the submission of infringement claims by Aerocar had to wait until the Curtiss design patents had been approved, a process that could take several years.

The potential sale of the Aerocar company after Curtiss's death had been hampered by the concerns of at least one potential buyer that the Curtiss Aerocar patents were not enforceable. A review of Aerocar's patents by attorneys in 1931 gave a similarly poor prognosis for patent protection under Aerocar construction patents, even suggesting that Aerocar may have itself infringed against several other prior patents. It is possible that Aerocar may have purchased these patents, since during the 1930s, Aerocar lawyers were confident enough to issue patent infringement notices against a number of trailer manufacturers. Infringement notices were issued against Harris and Schafer (Travel-Rite trailers) in October 1932; G. G. Coates (Coates & Brintnell) in June 1934; W. C. Lyter in June 1934; R. B. Springer of Hollywood, Florida, in February 1936; and George E. Tillitson in September 1938. Most cases were found in favor of Aerocar, although some were won only on appeal. By 1935, it must have become apparent to trailer manufacturers that Aerocar was not vigorously enforcing its patents, since on the cover of *Popular Homecraft* magazine of March/April 1935, readers were invited to build a travel trailer almost identical in appearance to a basic Aerocar.

Writing much later in 1937, Aerocar Florida president H. C. Genung's frustration with the patent situation boiled over. He wrote the following in a letter to the Doherty organization:

> The patent situation, thru [*sic*] lack of funds to conduct prosecution, etc., has been allowed to slip badly. Infringers are most brazen in all parts of the country. I really question whether our principal licenses will continue to pay royalties much longer on the ground that he no longer occupies a privileged position.
>
> —Letter from H. C. Genung to H. D. Williams, representative of Henry L. Doherty, May 18, 1937

The "An-X-Kar" manufactured by Eckland Bros. Company (1931)

It is interesting to speculate about whether the pursuit of the more relaxed Curtiss patent philosophy of just "going after the business" would have put the company in a stronger financial position following Curtiss's death. As it was, attempts to enforce Aerocar patents were poorly resourced, half-hearted, and ultimately achieved limited gains.

"A PLAGUE OF LOCUSTS"

200,000 trailers will swarm on the road this spring. Whether they betoken a New Way of Life or a plague of locusts is something that makers, taxpayers, hotelkeepers, and lawmakers are bitterly disputing.

—*Fortune* magazine, March 1937

In 1936, *Fortune* magazine had claimed that there were 75,000 tourist trailers in use in America and about 15,000 trailer camps. By 1937, the estimated number of trailers in use had swelled to 200,000. Trailer sales in 1936 were five times that of 1934, with some manufacturers producing as many trailers in the single year of 1936 as they had done in the five years previously. As mass production techniques were introduced, trailers priced between $300 to $600 were commonplace, with very few costing more than $1,000. Self-built trailers, using plans freely available in magazines, further lowered the cost of entry into the trailer market and provided severe competition for commercial manufacturers.

By 1938, the travel trailer industry boom was over, plagued less by locusts but more by overcapacity, nosediving demand, and poor-quality products. Trailer

The 1931 Tin Can Tourist Convention at Arcadia, Florida, was an early indication of how popular travel trailers would become in the 1930s. *Courtesy of the Tin Can Tourists*

manufacturers were burdened with unsold stock and silent plants. The long-term effects of the Great Depression included the emergence of trailer ghettos on the outskirts of many cities, which in turn led to concerns about health, sanitation, tax avoidance, and social cohesion. The term "gasoline gypsies" used in early Motorbungalo advertising had become derogatory—anyone living permanently in a trailer was clearly up to no good.

As part of President Roosevelt's New Deal, the Farm Security Administration agency was established in 1937 to alleviate rural poverty. It was one of the few large clients for trailers in the late 1930s, ordering thousands to house farmworkers close to their land. As the specter of war in Europe grew, auto, truck, and trailer manufacturers turned to war production of jeeps, tanks, weapons, and helmets. By the time of America's entry into World War II in 1941, the only trailers allowed to be manufactured were those used to house farmworkers or service personnel and their families close to military bases.

THE AEROCAR'S FINAL DAYS

With prices of up to $15,000 in a rapidly declining and price-sensitive market, the Aerocar was in a price league of its own. But the company stuck to its principles. In defiance of the trend to make trailers more affordable, in 1937 Aerocar was the first manufacturer to introduce air-conditioning into a travel trailer. The air-conditioning unit circulated 300 cubic feet of air per minute and included a ⅓-ton compressor and an 850-watt motor-driven generator to produce the required 110-volt AC power. It was an optional extra priced at $199.50.

But the bravado couldn't last. In Detroit, 1938 Aerocar advertisements were reduced to small silhouettes of four models including, with a sense of déjà vu back to the early 1920s, a utility trailer. From 1939 onward, the most numerous Aerocar advertisements were from private owners or dealers selling heavily discounted, used Aerocars. In Florida, the economic pain was felt just as severely. By March 1939, the Curtiss Aerocar Company advertised in the *Miami Herald* that it was offering "expert repairs and alterations on all makes of trailers at reasonable prices. Bendix and Warner electric brakes serviced."

Display Aerocars valued at $3,500 were being offered at $2,650, and $3,000 Aerocars were on sale at $1,650. As the tourist market in southern Florida dried up, some of the Aerocar club cars were sold off and acquired by local taxi companies.

Harry C. Genung had become president of Aerocar in Florida in 1938, with Harold Robinson moving to the role of vice president. In 1940, the company vacated its Coral Gables premises, downsizing to a small presence in the garage of the Miami Biltmore Hotel. One of the last vehicles to be sold by the company was its double-decker sightseeing bus, advertised for sale in February 1940. The Aerocar company ceased operations sometime during 1940 and never resumed after the war. As a sign of the times, the Aerocar plant at Coral Gables was taken over by Pan-American Airways in March 1941 for use as a radio plant and meteorology office.

A PLANE WITHOUT A PILOT

So what caused the eventual decline of the Aerocar? Automobile historian Roger B. White offers the following:

> By the mid-Thirties, it was obvious that Curtiss Aerocar was neither competitive in the burgeoning mid-price trailer field nor contributing to the democratization of the touring trailer concept except as an example to be imitated. . . . The company seemed to cling to the past; the links with Glenn Curtiss and his aircraft business were as strong as ever.
>
> —Roger B. White, *Automobile Quarterly* 32, no. 3 (1994)

There is certainly much truth to this, but there is more. From the outset, Aerocar had never sought to compete with lower-priced trailers. Even at the height of the trailer price wars, the Aerocar remained a premium product at a premium price. It was always a trailblazer rather than a democratizer. The primary cause of its downfall was its status as a premium product in a market that in the course of its life had become commoditized.

The fact that the company was unable in later years to convince customers of the value proposition offered by an Aerocar was partly due to economic reality, but also due to the lack of a post-Curtiss ambassador willing and able to navigate the company through difficult times. The company had plenty of ideas people at the start of the production cycle, and good engineers at the end, but no safe pair of hands in the middle to maintain operational and financial discipline in a crisis.

During its relatively short life, the Aerocar had been at first underestimated, then overegged, later exploited, and finally neglected. It had been pushed and pulled into a dozen products to suit different needs at different prices, in the process losing some of the beauty and soul given to it by its creator. A complex corporate structure involving patents, licenses, stock holdings, and multiple companies was put in place around the Aerocar, designed to sell a product in the thousands rather than the hundreds. Most of the Aerocar's owners were well meaning, but to them the Aerocar was always a sideline. When economic hurricanes hit, they didn't hesitate to leave the Aerocar in the garage while they tried to save their houses.

In different circumstances and under different management, the Aerocar might still be alive today, perhaps as an "Airstream of the South," built to exacting standards in Miami for a discerning clientele. But it was not to be, and in 1940, Aerocar ceased trading.

CHAPTER 7

THE CURTISS AEROCAR LEGACY

The genius of Mr. Curtiss will undoubtedly be felt in the auto design of the future. He has pointed to lightness, resilience, comfort, and safety at modest cost. His untimely death will not prevent the expansion of this new industry any more than the death of George Westinghouse prevented the development of the airbrake and signal industries.

Comparing the collection of land-based vehicles on the beach at Ormond-Daytona in 1905 with the road vehicles available to American consumers by 1940 gives some sense of the dramatic changes that took place in land transportation during the first half of the twentieth century. Curtiss's contribution toward this diversity was considerable. Although Curtiss did not live to witness the progress made in the 1930s, his personal achievements in advancing the comfort, safety, and flexibility of road transportation from his Wind Wagon of 1906 to his Aerocar of 1928 left a lasting mark. Much has been written about his achievements in the air, but his inventions in the field of ground transportation have been underreported and underrecognized.

This early image of the G. H. Curtiss Manufacturing Company's factory at Hammondsport gives an indication of the breadth of vehicles over which Curtiss would have an influence during his life (1907). It shows an automobile, motorcycles, a bicycle, and (right) the Wind Wagon.

HOW SAFE WAS THE AEROCAR?

Newspaper reports of accidents involving Aerocars were rare. Some incidents were the result of the Aerocar's popularity. In 1928, an automobile ran into a crowd inspecting a parked Aerocar at Palm Beach, Florida. In 1929, an Enna Jettick Aerocar traveling to Alexandria, Indiana, for a footwear demonstration was "partially disabled" as a result of being run into by another motor vehicle. In 1931, an Aerocar gasoline stove caught fire in Milton, Pennsylvania, with the owner suffering serious burns to his hands. In 1938, two members of a baseball team were injured as the "Hialeah aero car" they were traveling in met with an accident near Key West in Florida. The most serious known Aerocar accident happened in 1932 to the unfortunate Ernest R. Teague, who was killed near Okeechobee, Florida, as a result of head injuries sustained after putting his head out of an Aerocar window due to feelings of travel sickness and being hit by an oncoming truck.

The towing safety offered by the Aero Coupler was undoubted, but the risks to passengers traveling inside an Aerocar at speed were clearly high, especially when considering external impacts to a vehicle with a light frame and next to no passenger safety features such as seat belts. Aerocar was not alone in prioritizing aerodynamics over safety. Most automobile manufacturers of the 1930s were more interested in the streamlined features of their vehicles, which, as it turned out, reduced rather than enhanced visibility. Automobile crash testing did not begin in any serious way until the 1950s.

We do know that some potential Aerocar customers did not proceed with their orders due to safety concerns. In 1929, well-known journalist Arthur Brisbane reportedly canceled his Aerocar order because he was concerned about his children traveling in the trailer. He was worried about the lack of safety glass (which was in fact an Aerocar option) and the general lightness of the frame. In 1932, General Motors conducted a monthlong trial of an Aerocoach for possible use as a Greyhound Bus but returned the vehicle without comment and without placing an order. It's possible that GM, a pioneer in automobile safety, had similar concerns.

Even today, the act of carrying passengers inside travel trailers at speed carries a high degree of risk. Travel trailers do not have the same occupant protection levels as automobiles. Although still permitted by some states, the safety advice is simple—don't do it.

HOW MANY AEROCARS WERE BUILT?

The total number of Aerocars built is a mystery. Company manufacturing records were either poor or have been lost over time. The large number of different models made at two different factories complicates matters further. Images handed down to the Glenn H. Curtiss Museum have serial numbers on the back, but it's not always clear if these numbers relate to models produced or are image numbers of the photographer. Some of the few Aerocars still in existence have serial numbers, but too few remain to be able to draw any conclusions about numbers built.

We do know that up to August 1930, the Aerocar Company of Detroit had manufactured 173 Aerocars (including eighty-four deluxe A-20 touring models and thirty-one commercial C-20 models) and, by 1932, had reached the production serial number 550, but no further details remain.

An educated guess as to the total number of Aerocars built would be about a thousand across both factories and of all model types.

How are we then to assess the legacy of the Curtiss Aerocar? Here's a summary of its key achievements.

THE FIRST AMERICAN FIFTH-WHEELED TRAILER BUILT FOR LEISURE

The Curtiss Camp Car of 1919 that served as the Adams Motorbungalo prototype was, as far as we know, America's first fifth-wheeler used for leisure purposes. Prior to that date, fifth-wheel trailers had been used in the country only to haul goods. It's likely that Curtiss adopted the fifth-wheel format on the basis of firsthand observation of planes being towed between factory and airfield. We know that he also considered an integrated house car concept, but weight and flexibility steered Curtiss toward the trailer configuration when going into production with the Motorbungalo.

Although abandoned for Motorbungalo production models in favor of the simpler ball-and-socket hitch, the fifth wheel was used again in the Trailer Bus of 1922 and the Aerocar of 1928. By the late 1920s, the stability benefits of the fifth-wheel format at the high speeds achievable by the Aerocar would have become apparent. Although towing dynamics were not formally studied until the 1950s, Curtiss was a true pioneer in understanding and applying the important safety benefits of fifth-wheelers.

These pioneering initiatives can be fully appreciated today in the context of the recreational fifth-wheel market. They are one of the most important sections of the leisure vehicle sector and one of the safest. In 2021 alone, over 107,000 fifth-wheel leisure trailers were shipped from US manufacturers to dealers.

THE FIRST FULLY ENCLOSED TRAILER IN AMERICA

Curtiss's recreational vehicles were instrumental in the transformation of the American auto trailer from crude camping vehicles to sophisticated, solid-walled trailers. The Adams Motorbungalo demonstrated that a camping trailer could be both luxurious and comfortable while being flexible and mobile. The Motorbungalo prototype also served as the preprototype of the Aerocar.

The Aerocar was, as far as we know, the first solid-walled, solid-roofed trailer to be manufactured in America, starting in 1928, preceding Covered Wagon and others by at least a year. Its arrival heralded new levels of comfort for leisure travelers, allowing them to do away with "roughing it" in tents or canvas-based trailers and instead to enjoy fully enclosed comfort with beds, kitchens, and bathrooms.

Curtiss, strongly encouraged by Carl G. Fisher, was also a pioneer in the development of multipurpose trailers that could carry people, goods, or animals. The lightness and open design of his trailer allowed for a wide variety of payloads and facilitated a new form of interaction between businesses and their customers in the form of the mobile showroom.

THE FIRST PRODUCTION MONOCOQUE RECREATIONAL VEHICLE

The aircraft principles used in the construction of the Aerocar led to the creation of a road-going monocoque chassis. When automobile manufacturing was divided into the separate empires of chassis making and body building, the result was often a heavy and unwieldy vehicle. Curtiss did away with the chassis altogether and instead created a single, aircraft-inspired strut, wire, and fabric frame that dramatically reduced weight while retaining strength. Although Charles Louvet of France built recreational vehicles on similar aircraft-inspired lines from about 1923, they never went into production. Curtiss was the first to use this technique in production recreational vehicles.

THE FIRST PNEUMATIC COUPLER

The Aero Coupler was the first-known pneumatic-coupling device. Not only was it used in a relatively stable fifth-wheel configuration, but its shock-absorbing features improved ride comfort for passengers. This made long journeys more comfortable and safer. As Fisher had eloquently said, the Aero Coupler was "as simple as an old shoe," but it served its purpose admirably.

So why are Aero Couplers not in use today? Simply put, it was because of progress in materials and suspension technology. As trailers became heavier and faster, the physical forces acting on trailer couplings grew. New metal alloys, stronger rubber, tensile springs, and efficient hydraulic mechanisms were required to deal with these forces, which were well beyond the capability of an inflated aircraft tire.

THE FIRST LONG-DISTANCE RECREATIONAL VEHICLE

The combined effect of its fifth-wheel configuration, monocoque chassis, solid walls, and Aero Coupler meant that long journeys could be undertaken safely and comfortably in an Aerocar. A reasonable amount of creature comforts on board increased the possibility of longer stays away from home. The arrival of the Aerocar on the roads of America created competition for the railroads and later the airlines. Meanwhile, alongside the country's roads, new tourism facilities grew in the form of a network of trailer parks, including in some of the nation's more remote areas such as the new national parks. Even if a customer could not afford an Aerocar, its many admirers were given tangible evidence that the travel trailer was a viable form of transport not just for weekends but for months, and not just locally but nationally. It was the first truly long-distance touring vehicle in America.

Sample Aerocar brochure covers illustrate the increasing breadth and sophistication of the Aerocar during its production.

1929

1929

1929

1933

1936

1938

AUTOMOBILE DESIGN INNOVATIONS

Curtiss did not live long enough to influence automobile design in the ways that he would have perhaps wished. His diagnosis of "auto-intoxication" led to the development of the Aerocar, but it also led to a radically streamlined motor vehicle in 1930, the Curtiss Motor Vehicle, which was years ahead of its time. Its front-wheel drive and use of Aero Coupler–like suspension to reduce noise and vibration were revolutionary. Had Curtiss lived longer, it may have become the Aerocar motor home.

The streamlining trend of automobiles and recreational vehicles that began in 1930 was sometimes followed for appearance's sake rather than to improve the aerodynamics of fast-traveling vehicles. But when the Aerocar could travel at speeds of up to 70 mph (113 km/h), streamlining became a much-needed safety feature. Rounded roof, front and rear ends, a bird-beak hitch, a low center of gravity, and a long wheelbase were important both for streamlining and safety. Suspension and articulation improvements further added to the comfort and safety of his vehicles.

Curtiss's contribution to automobile design was not just in the realization of his own ideas but also in his articles on the topic. These encouraged others to think about how to build cars better. In stimulating debate about automobile design efficiency, he helped make automobiles of many makes lighter and safer.

AN EARLY WELFARE CAPITALIST

During Curtiss's flying days, we see glimpses of his business philosophy. His establishment of flying schools and airfields around the country is an early example of his broad vision for a fledgling airline industry, which would not thrive without pilots and places to land. He was always looking to the future and considered at some depth what a new industry would need to become successful. Curtiss's move to Florida shows this big-picture approach in action.

Although "welfare capitalism" is a grand term for Curtiss's business philosophy in Florida, it does capture something of the spirit that drove his activities there. Yes, he was a real estate developer, and yes, he made money from it, at least in the early 1920s. But he also built new communities, donated land and water rights, provided jobs for old colleagues and locals, made efforts to support the indigenous Seminole tribe, and acted as an ambassador for South Florida. He was less interested in beachfront real estate and more in back-country lots suitable for farming. Agriculture and small industry would help South Florida become a year-round destination, not just a winter escape. Together with James Bright, he experimented with new crops, fruit, and animals and sought to farm sustainably.

The retrenchment of his aircraft factories after World War I and the resultant heavy job losses clearly affected Curtiss. The fact that so many of his former colleagues followed him from Hammondsport to Hialeah is testament to the high regard in which he was held. He looked after them as best he could and was repaid with high-quality workmanship in the Aerocar that continued long after his death. His philanthropic spirit may have exasperated his business partners, but his wishes were nearly always respected. Toward the end of his life, his trust and generosity may have contributed to business problems, but it's unlikely that he would have changed the habits of a lifetime.

One of Curtiss's greatest achievements is that the Curtiss Aerocar is still known to many Americans today. Perhaps it would be a fitting tribute if, by the centenary of its invention in 2028, the relevant authorities in, say, Hammondsport or Miami could restore an Aerocar or build a modern Aerocar replica, not for viewing in a museum, where it does not really belong, but on the open road for the use of visitors who might marvel at its aerodynamic shape, its spacious interior, and its comfort.

THE RELUCTANT VISIONARY

Curtiss was a quiet man. His newspaper interviews were generally short, as were his speeches. But what he said was generally worth listening to. Curtiss's predictions were not always right, but many were. He predicted the first man on the moon forty years before it happened. His earthlier predictions of long-distance passenger flights also came to pass, although his hopes for duck-shooting from aircraft failed to take off.

His prediction of a boom in aerodynamic automobile design in 1929 presaged a decade of streamlined road vehicles. His last road vehicle of 1929 was ahead of its time and a decade ahead of other designers with similar ideas.

Curtiss was a reluctant visionary because he did not speak out in order to achieve fame. When he did make a public comment, it was because he wanted his engineering and design contemporaries to think about what they were doing and how they could do things better. Curtiss wanted to share his ideas for the wider benefit of the industries he worked in. His patents existed to protect the inventor, not to prevent innovation.

TEAM CURTISS

It is convenient but inaccurate to attribute the achievements of Curtiss's organizations and the machines bearing his name to the man and the man alone. Curtiss himself would not have wanted this to be his legacy. He was a team player who inspired, educated, and supported those around him. He understood his own weaknesses and found colleagues or partners to compensate for them. He was a great delegator and happy to give credit to others where it was due. He was surrounded by some of the best engineers, mechanics, salesmen, and financiers in the business. The Aerocar would never have come into being without this team-based approach. Those who formed part of the Aerocar team included G. Carl Adams, Curtiss's half brother, who would do anything asked of him by Curtiss and who worked tirelessly across many of Curtiss's organizations over two decades; Hugh Robinson, a trusted pilot, driver, and mechanic who probably built the first Aerocars, managed the Florida factory for many years after Curtiss's death, and, along with his son Harold, came up with a number of Aerocar-based vehicles of their own; Henry Kleckler, an engineer who designed and built Curtiss engines and built the motorized

From a fruit and vegetable transporter came many things (1929).

Aerocar prototype; Carl Fisher, who saw the potential of the Aerocar and sold its potential to the automobile industry; C. M. Keys, a financier and friend who believed in Curtiss, bought him out of the airplane-manufacturing industry, made him a millionaire and helped him secure financing to realize his other projects; and loyal friends Harry Genung and H. Sayre Wheeler, who kept the business ticking over and who took turns running Aerocar in Florida after Curtiss's death. There were many more.

Why did Curtiss command such loyalty? Early naval aviator Theodore "Spuds" Ellyson gives us a clue when describing a series of seaplane experiments in San Diego in 1911:

> According to Spuds Ellyson, no less than fifty changes were made in the course of these experiments. "Those of us who did not know Mr. Curtiss well," he said in retrospect, "wondered that he did not give up in despair. Since that time we have learned that anything he says he can do, he always accomplishes, as he always works the problem out in his mind before making any statement. . . . You see it was not Curtiss, the genius and inventor, whom we knew. It was 'G. H.,' a comrade and chum, who made us feel that we were all working together, and that our ideas and advice were really of some value. It was never a case of 'do this' or 'do that,' to his amateur or to his regular mechanics, but always, 'What do you think of making this change?' He was always willing to listen to any argument but generally managed to convince you that his plan was best. I could write volumes on Curtiss, the man."
>
> —From *Glenn Curtiss, Pioneer of Flight*, by C. R. Roseberry

THE AEROCAR LEGACY

In 1910, Curtiss was said to have likened an aeroplane to a "monster violin." If his planes were monster violins, the Aerocar could perhaps be likened to a monster cello. It was a finely tuned instrument that soothed its passengers, insulating them from a noisy, harsh world outside. By contrast, the Aerocar's travel trailer competition during the 1930s would have been more at home in the orchestra's percussion section.

If the Aerocar had been just a travel trailer, its legacy in terms of its impact on the recreational vehicle industry of America would have been powerful enough. Thousands of tourists were able to explore the country's national parks, lakes, forests, and wilderness areas in mobile comfort and safety thanks to the Aerocar.

But the Aerocar's greatest achievement is as a multipurpose road vehicle. Before the Aerocar, the powered vehicles seen on America's roads came in three basic forms: motorcycles, automobiles, and trucks. After the Aerocar, there was also the fifth-wheeler in all its forms: club cars, ground transportation cars, commuting cars, mobile salesrooms, display cars, ambulance trailers, transport for the disabled, semitrailer passenger buses, horse trailers, garbage trailers, mobile speaking platforms, and even mobile churches—not in large numbers, but enough to get noticed by the automobile industry and by consumers.

The discipline and attention to detail of the aircraft builder were transferred from the skies to the road. Lightweight and vibration-absorbing construction methods used in the Aerocar sent a message to automobile and truck manufacturers that they needed to lift their game. This they eventually did, opening up new possibilities and new levels of speed, comfort, and safety in road transportation of all kinds.

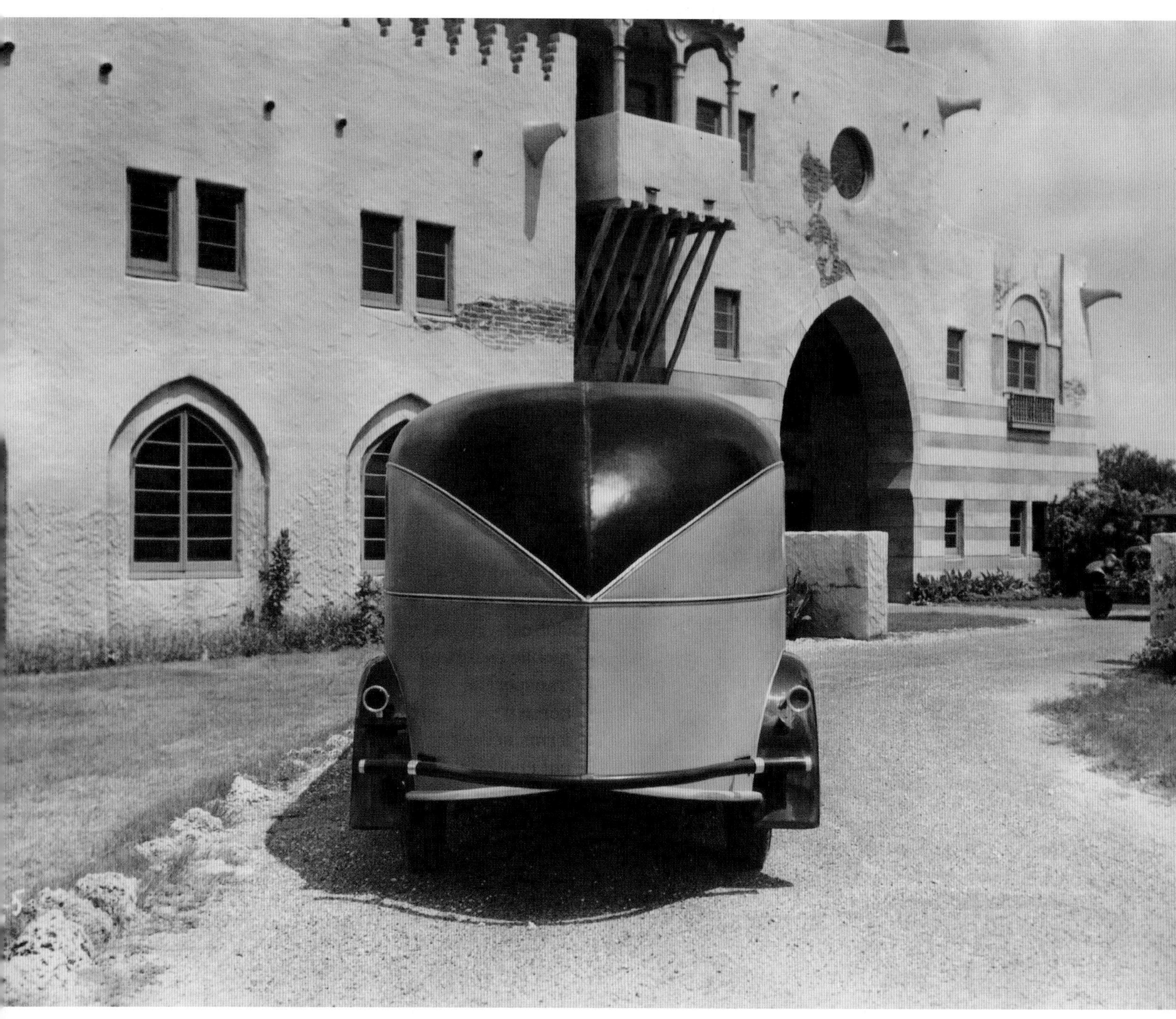

APPENDIX A

AEROCAR-RELATED PATENT EVOLUTION

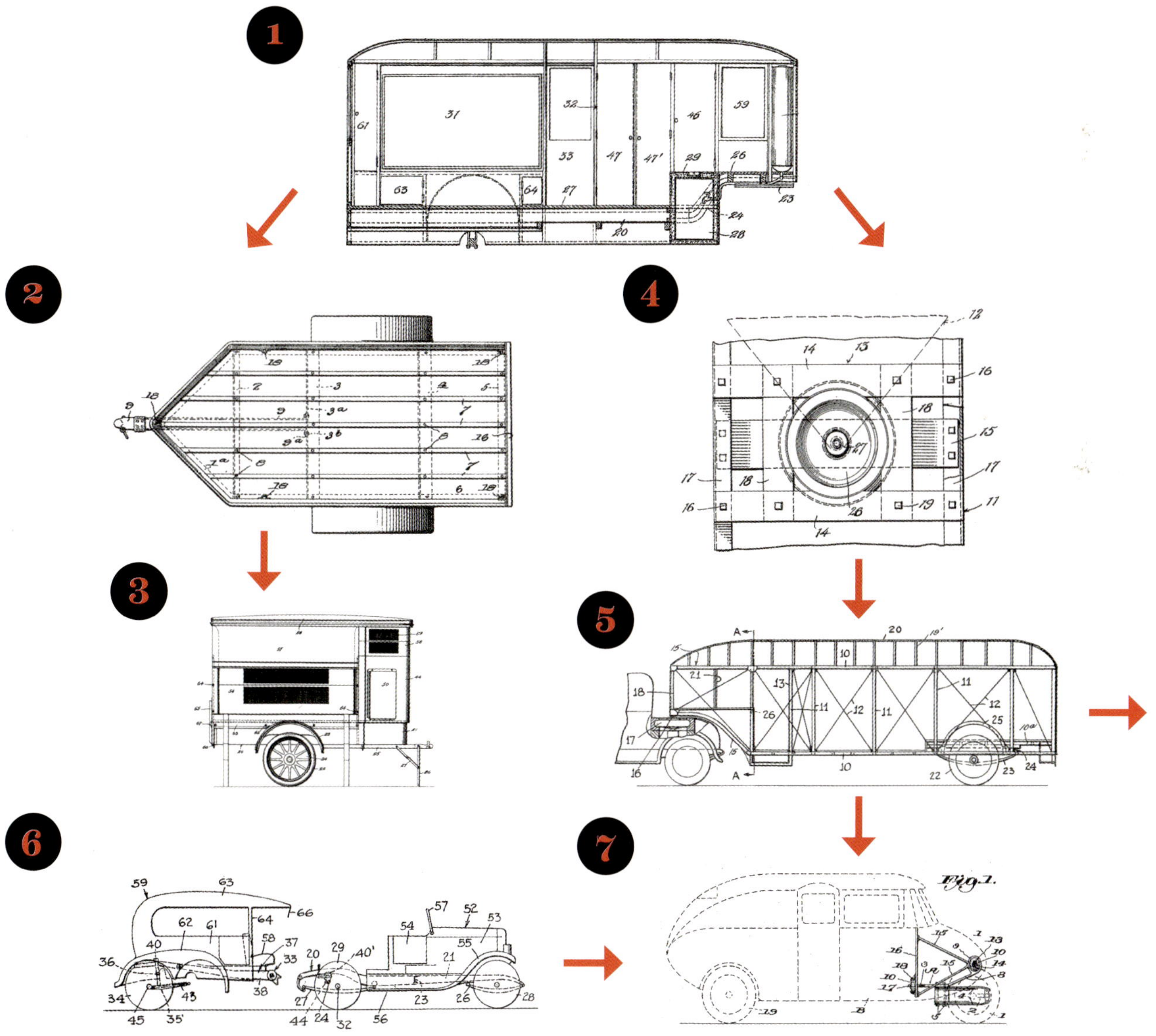

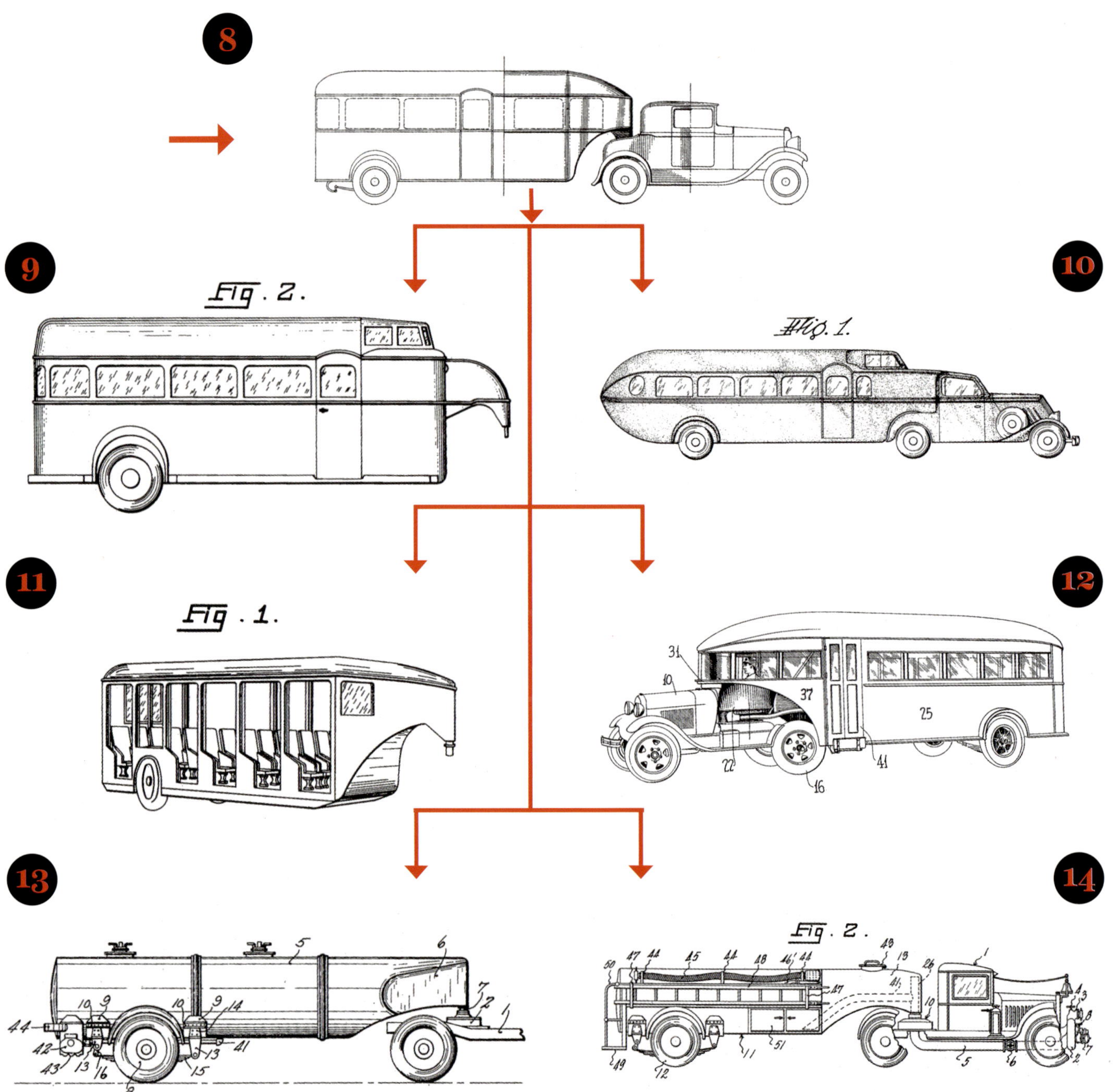
8
9
Fig. 2.
10
11
Fig. 1.
12
13
14
Fig. 2.

1. Patent number US 1,437,172, G. H. Curtiss, Camp Car

Application filed April 28, 1921; patented November 28, 1922

Key features: Multipurpose camp car with three beds (two of which were folding) and canopies, cooking, refrigerator and dual-access cabinets (from inside or outside), raised floor, and step. Overhanging and tapered front end with coupling attached to tractor (coupling details to be in a "separate application" not known to be submitted).

Comment: This is the first known design for a fifth-wheel trailer in the US intended for leisure. Since a separate patent was not submitted for the coupling around that time, it is likely that an existing semitrailer fifth-wheel coupling was used. This camp car was used privately by Curtiss but was deemed to be too large and heavy for commercial production.

2. Patent number CA 222875, George C. Adams, Trailer

Application filed April 28, 1921; patent issued August 22, 1922. Canadian patent, no US patent known.

Key features: Sprung-steel trailer for use with automobiles. V-shaped front for strength and to permit coupling close to automobile without interfering with turning. Strengthening iron floor rabbets, cross braces, and pressed-steel sides.

Comment: Designed by Adams for commercial work, this trailer was to become the chassis for the Adams Motorbungalo. V-shaped front increased towing stability and suggests Curtiss's input.

3. Patent number CA 221009, George C. Adams and John A. Methot, Trailer

Application filed March 22, 1921; patent issued July 28, 1922. Canadian patent, no US patent known.

Key features: Folding camping trailer with V front, adjustable top, and side canopies that lifted as folding beds were lowered. Roof rests on internal cabinets when lowered. Folding rear step. All camping fixtures detachable from trailer bed for use as a commercial trailer.

Comment: A camping trailer that came to be known as the Adams Motorbungalo. Shorter and lighter than the Curtiss camp car and with conventional hitch for multivehicle use, but otherwise with many identical features. Curtiss known to be principal designer. John A. Methot worked for Maytoe-Perry Body Designing Co. of Newark, New York, and may have contributed to the body's design.

4. Patent number US 1,916,967, G. H. Curtiss, Flexible Coupling for Vehicular Structures

Application filed June 8, 1928; patent approved July 4, 1933

Key features: Flexible coupling between power vehicle (road, rail, air, or water) and trailer. Pneumatic tube coupling absorbs shocks in all directions. Tire-like rubber tube is held in frame on rear of tractor. Aeroplane tire suggested; kingpin of trailer is inserted into center of tire. Rubber "doughnuts" or compression spring suggested under tire.

Comment: The Curtiss Aero Coupler patent

5. Patent number US 1,880,844, G. H. Curtiss, Road Vehicle Body Structure

Application filed June 8, 1928; patent approved October 4, 1932

Key features: Box girder, chassisless trailer for merchandise and passengers. Low-cost, lightweight, more comfortable, and safer construction than conventional trailers. Uses tensioned piano wire and struts (or steel tubes) with light weatherproof covering for sides and roof. Structure will not collapse if turned over. Designed for use with Aero Coupler (*see above*) but works with any coupling.

Comment: The monocoque structure used in aircraft is applied by Curtiss to road trailers. The commercial Aerocar.

6. Patent number US 1,682,324, G. H. Curtiss, Automotive Vehicle

Application filed August 8, 1925; patent issued August 28, 1928

Key features: An automobile designed for improved ride comfort of occupants. Chassis divided into two parts, the rear section pivoting over the front section and attached via a "yielding shock absorber" consisting of a spring, rods, and rubber discs. Single-point front connection of rear section improves absorption of lateral shocks.

Comment: An invention that shows Curtiss's preoccupation with improving ride comfort by separating tractor and trailer and using triangular rather than four-point chassis connections. The forerunner of the Curtiss Motor Vehicle of 1929.

7. Patent number 1,948,744, G. H. Curtiss, Motor Vehicle

Application filed July 9, 1929; patent approved February 27, 1934

Key features: A single-bodied, front-wheel-drive motor vehicle that reduces engine and road shock and vibration to passengers by separating the body from the propulsion and steering mechanism. The body and engine are connected by triangular "yielding coupling members" incorporating up to three vertically mounted pneumatic couplers similar to the Aero Coupler.

Comment: An attempt to design a "one-box" version of the Aerocar. Only one prototype is believed to have been built shortly before Curtiss's death by Curtiss engineer Henry Kleckler. The earliest known streamlined "house car" in the US.

8. Patent number USD 85,815, Lena P. Curtiss, Design for a Tow Car and Trailer Combination

Patented December 22, 1931

Key features: A patent to protect the exterior design of the Aerocar, in particular the V-shaped trailer prow, converging roof, and concave curve of lower front end

Comment: Submitted by Curtiss's widow, Lena P. Curtiss, after his death

9. Patent number USD 90,169, H. H. Robinson, Trailer

Application filed March 2, 1933; patent approved June 20, 1933

Key features: Design patent showing modified Aerocar with upper-level front cockpit for passenger viewing purposes

Comment: Submitted by Curtiss Aerocar Company engineer Harold H. Robinson after Curtiss's death

10. Patent number USD 90,925, H. H. Robinson, Combination Tractor and Trailer

Application filed March 27, 1933; patent approved October 24, 1933

Key features: Design patent showing integrated tractor and trailer with trailer cockpit and streamlined rear

Comment: Submitted by Curtiss Aerocar Company engineer Harold H. Robinson after Curtiss's death

11. Patent number USD 91,965, H. H. Robinson, Open Trailer

Application filed March 27, 1933; patent approved April 10, 1934

Key features: Design patent showing open-sided passenger bus trailer. Another Robinson patent (USD 94,096, patented on December 18, 1934) proposed an open-top passenger bus trailer.

Comment: Submitted by Curtiss Aerocar Company engineer Harold H. Robinson after Curtiss's death

12. Patent number US 1,980,613, G. H. Curtiss, Motor Vehicle

Application filed April 1, 1931; patent approved November 13, 1934

Key features: A tractor-and-trailer combination that provides an interconnection between driver and passengers for the purpose of collecting fares and other uses. Fifth wheel is located farther forward on tractor for better load distribution and, with the rear axle of the tractor "pushing" rather than "pulling" the trailer, increased stability.

Comment: The Aerocoach, a passenger bus version of the Aerocar. Submitted by Curtiss's widow, Lena P. Curtiss, after his death.

13. Patent number US 2,036,607, H. H. Robinson, Tank Trailer

Application filed July 11, 1933; patent approved April 7, 1936

Key features: Liquid-carrying trailer for gasoline, oil, water, etc., using Aerocoupler hitch to absorb shocks

Comment: The Aerotank patent. Submitted by Curtiss Aerocar Company engineer Harold H. Robinson after Curtiss's death. One of several commercial applications using the Aerocar design and Aero Coupler hitch.

14. Patent number USD 1,982,052, D. E. Hennessy, Semitrailer

Application filed August 7, 1933; patent approved November 27, 1934

Key features: A firefighting semitrailer with self-contained water pump, using the Aerocar design

Comment: Submitted by Curtiss Aerocar Company employee Daniel E. Hennessy after Curtiss's death

APPENDIX B

SAMPLE ADAMS MOTORBUNGALO AND AEROCAR ADVERTISING

Gypsie Life—Modernized!

ANNOUNCING the camp-de-luxe, designed to meet the exacting wants of the man who demands the maximum pleasures of camping.

Imagine a nine by six room, electrically-lighted, containing a lounge, kitchenette, food lockers, ice-chest, full-length clothes closets, running water available at all times, together with two double beds which can be opened out in less than two minutes!

And add to these, exceptional lightness—accessible equipment--attachability to any make of car.

These are only a few of the features that make this auto-bungalow the only practicable, comfortable and luxurious vehicle for touring and camping.

For particulars address

G. Carl Adams, Farmingdale, Long Island, New York

The first known advertisement for the Adams "camp-de-luxe" camping trailer in *Outers' Recreation*, July 1920

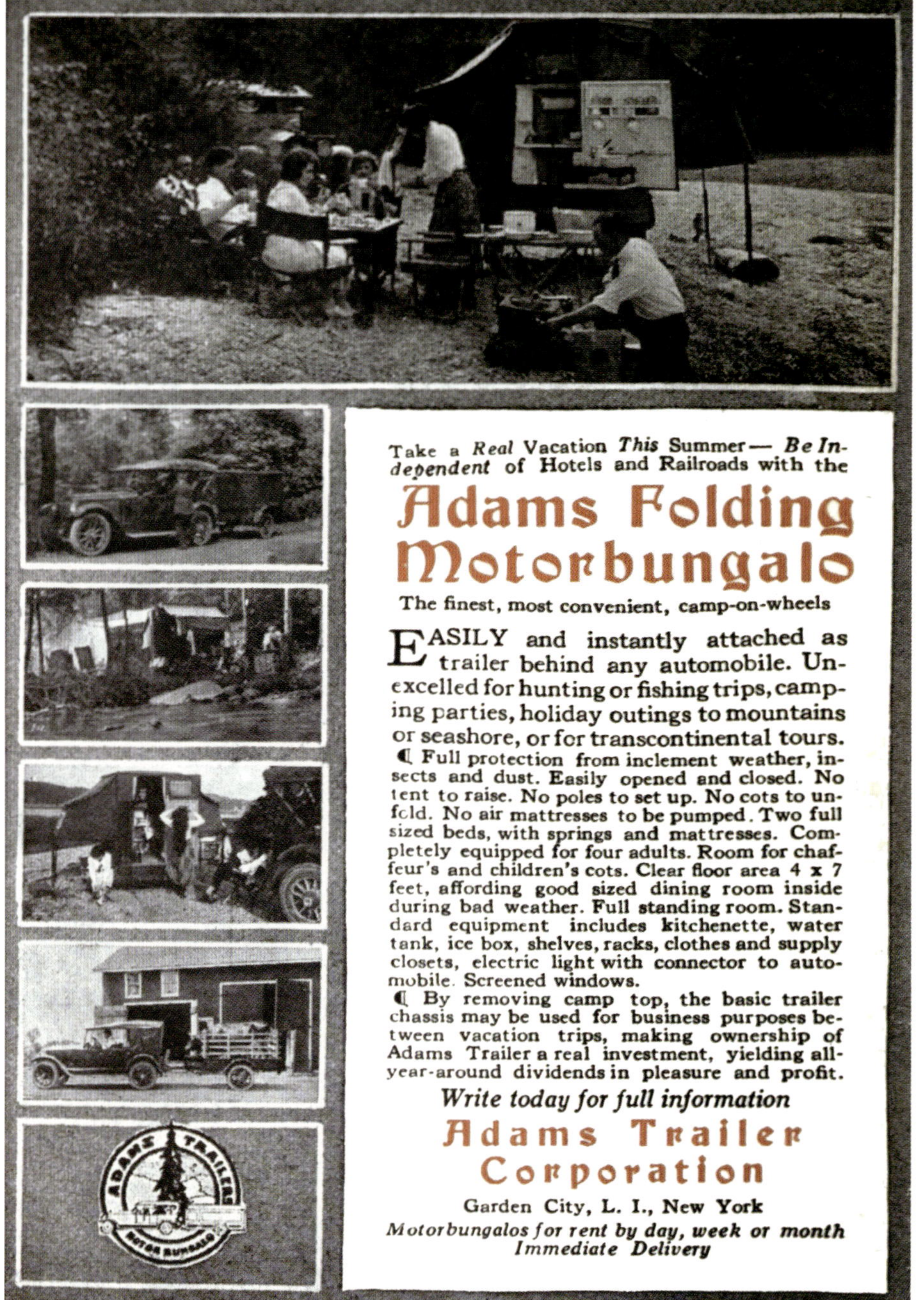

The first known advertisement for the "Adams Folding Motorbungalo" in *Roycroft* Magazine, October 1920

An Adams Folding Motorbungalo advertisement in *Field and Stream*, July 1921

An Aerocar advertisement in the *Baltimore Sun*, October 13, 1929

An Aerocar advertisement in the *Miami News*, February 2, 1930

A New Type of Road Transportation

THE wonderful riding qualities of the Curtiss Aerocar will appeal to all who travel, as well as to business officials and executives, professional men. The Curtiss Aerocar will accommodate as many as 14 people comfortably. It is easy to maneuver in traffic, light, swift and economical of operation. Already the Curtiss Aerocar is being adapted to a wide variety of uses, as evidenced by the abundant testimony shown below.

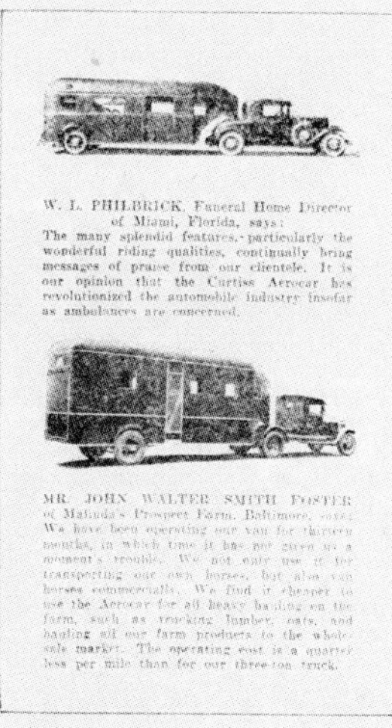

Curtiss Aerocar Co., of Florida, Inc.

OPA-LOCKA, (*Miami*) FLORIDA

An advertisement by the Curtiss Aerocar Co., of Florida, Inc., featuring a range of Aerocars, including an ambulance, airline transport car, horse transport truck, and garbage truck, in the *Miami Daily News*, May 25, 1930

An advertisement by the Miami Biltmore Hotel including a photograph of its fleet of fourteen Aerocars in the *Central New Jersey Home News*, January 22, 1933

An advertisement by the Curtiss Aerocar Co. Inc., Coral Gables, Florida, in *Modern Packaging*, March 1934

An advertisement by the Aerocar Company of Detroit in *Sales Management*, October 10, 1937

An advertisement by the Aerocar Company of Detroit in *Electrical Merchandising*, August 1938

APPENDIX C

AEROCAR MODELS, SPECIFICATIONS, AND PRICES

Aerocar models, specifications, and prices varied significantly depending on place and year of manufacture. In broad terms, Aerocars made at Opa-locka and, later, Coral Gables were more expensive and custom built, while models built in Detroit were lower-cost production models.

During the life of the Aerocar, a mind-boggling number of models were produced. The following list contains details of thirty-seven models but is incomplete.

OPA-LOCKA MODELS (1929)

Ten "deluxe" models were initially offered in Florida, as part of the Series 61 or Series 71 range:

Model 61 (trailer only), $2,500
Model 61-A (ambulance), $3,000
Model 61-C (camping), $2,600
Model 61-G (commercial/garbage truck), $2,850
Model 61-H (4-horse trailer), $3,500
Model 61-HS (tall 4-horse trailer), $3,750
Model 61-O (observation windows all round), $2,750
Model 61-OS (observation with reclining seats at rear), $2,900
Model 61-P (12-passenger transport bus), $2,500
Model 71-S (streamline club car), $3,000

September 1929 options from the Florida factory included leather upholstered beds ($145), carpets ($40), washbasin ($35 and upward), gas stove ($30), marine-type lavatory ($110), speaking tube ($27.50), radio ($200), refrigerator ($100), silk curtains ($5.50 each), cigar lighter ($9), and gun rack ($4). Customers could also specify a rear speedometer, barometer, altimeter, thermometer, and clock.

DETROIT MODELS (1929)

Four "standard" models were initially offered in Detroit, using a simpler A, B, C, or D naming system followed by "20" for the Aerocar's length:

Model C-20 (Standard Commercial Car), $1,000, 20 ft. long, 1,750 pounds
Model B-20 (Standard School Bus), $1,200, 20 ft. long, 1,850 pounds
Model A-20 (Standard Tourist Car), $1,500, 20 ft. long, 1,900 pounds
Model D-20 (Standard Passenger Bus), $1,500, 20 ft. long, 1,850 pounds
No options were mentioned in early Detroit brochures.

OPA-LOCKA MODELS (1932)

In 1932 a "standard model" (161-C) was added to the Florida Aerocar deluxe series. The standard Model 161-C was 19 ft., 4 in. long and 5 ft., 10 in. wide and had four berths, leatherette finish, bumperettes, Watson stabilators, lino floor, "air cushion coupler," tripod legs, gas stove, toilet, fridge, and roller shades. It came in four variants:

Model 161-C chassis only, $1,985
Model 161-C with kitchen, $2,365
Model 161-C with kitchen and 2 berths, $2,645
Model 161-C with kitchen and 4 berths, $2,840

CORAL GABLES MODELS (1933)

Following the company's move to Coral Gables in July 1933, a number of new models were introduced in Florida:

Model 161-B. "Custom features for a fixed price." Touring for six, sleeping for four. Four chairs convertible to two beds, 2 Pullman uppers, table and seating up front, galley, toilet, shower, gas heater, hot water, wardrobes, 100- and 6-volt lighting and radio, telephone between car and power unit, adjustable windows with shatterproof glass, Bendix vacuum-operated brakes, Lovejoy shock absorbers, Aerocoupler. Weight 4,700 pounds; 23 ft., 2 in. long and 7 ft., 6 in. wide. $5,235.

Model 100-JCA. "Fully-equipped mobile home." Two bedrooms, kitchen, and bath. As per 161-B except shorter wheelbase, "semi-airplane construction," studio instead of Pullman beds, fixed windows, bamboo chairs, radiant gas heating, Warner electric brakes, Aerocoupler. Weight 3,500 pounds; 21 ft., 8 in. long and 7 ft., 6 in. wide. $3,600.

Model 161-SP ($4,500–$6,500). Includes chauffeur's compartment at rear

Model 161-E ($4,050–$4,250). As per Model 161-B but with marine flushing toilet, stainless-steel galley and stove hood, commodious cabinets.

Model 161-BS ($4,500–$6,000). Four chairs convertible to single beds, upper berths.

Model 161-BPC ($6,000–$8,500). As per Model 161-B with observation cockpit.

Further models (e.g., Model 150-C, Model 151-C, and Model 162-C) were added but no specifications given in brochures. Four bus models were added:

Model 161-P ($3,850). 14-seater bus.

Model 61-P ($3,700). 12-seater bus plus baggage compartment.

Model 261-P ($4,400). 23-seater bus.

Model 261-PC ($5,200). 23-seater bus plus observation cockpit.

DETROIT MODELS (MID-1930S)

New leisure Aerocar models from Detroit in the mid-1930s were called "Land Yachts" and given names such as "the Sportsman." A 16-foot-long model was added:

Model A-16 Land Yacht "the Sportsman" ($1,300). 16 ft. long, 1,700 pounds, forward galley.

Model A-20 Land Yacht "the Cruiser" ($1,650). 20 ft. long, 1,850 pounds, rear partitioned galley, divan seats.

Model A-20 Land Yacht "the Tourist" ($1,650). 20 ft. long, 1,850 pounds, rear partitioned galley, wicker seats.

A shorter model was also added to Detroit's commercial range:

Model C-16 ($1,000). 16 ft. long, 1,500 pounds.

Model C-20 ($1,150). 20 ft. long, 1,650 pounds.

Model C-22 (price on request). 22 ft. long, 2,000 pounds and up.

LATE 1930S DEVELOPMENTS

In a first for the travel trailer industry, air-conditioning was made available as an option in December 1937 at a cost of $199.50. It was a Pleasantaire unit with a 1/3-ton compressor and was powered by an 850-watt, motor-driven generator producing 110-volt a.c. The generator could also power an electric fridge.

In about 1938, a steel-bodied Aerocar was offered as an option in Detroit. The steel bodies were probably subcontracted to the Novi Products Company, whose Detroit address was shared with Aerocar.

By 1939, Aerocar display models were being offered at significantly reduced prices due to oversupply in the travel trailer market. In March 1939, the Coral Gables factory advertised three Aerocar display models for $1,650 (originally $3,000), $2,650 (originally $3,500), and $3,350 (originally $5,235).

In February 1940, the Aerocar double-decker bus was advertised for sale at an undisclosed price. Inquiries were to be made at the Biltmore Hotel Garage in Miami.

The last known commercial sale by Aerocar in Florida was in September 1940. The Curtiss Aerocar Co., Miami Biltmore Hotel garage, advertised a Pontiac 1938 sedan, a "sacrifice at $500."

APPENDIX D

PRIVATE AEROCAR GALLERY

Philip K. Wrigley, Chicago

W.K. Vanderbilt, Long Island

Arthur K. Browne, California

John L. Senior, New York

Henry L. Doherty, New York

Charles A. Munn, Palm Beach

W.G. Potts, Chicago

G. MacCulloch Miller, Long Island

Marquis George MacDonald, Princeton, New Jersey

C.W. Bell, Boston

APPENDIX E

BUSINESS AEROCAR GALLERY

Ogden Candy (1928)

Peoria Casket Company (1929)

Southwest Air Fast Express (1929)

Transcontinental Air Transport (1929)

Congressman Skipper (1930s)

J. E. Burke Co (1930s)

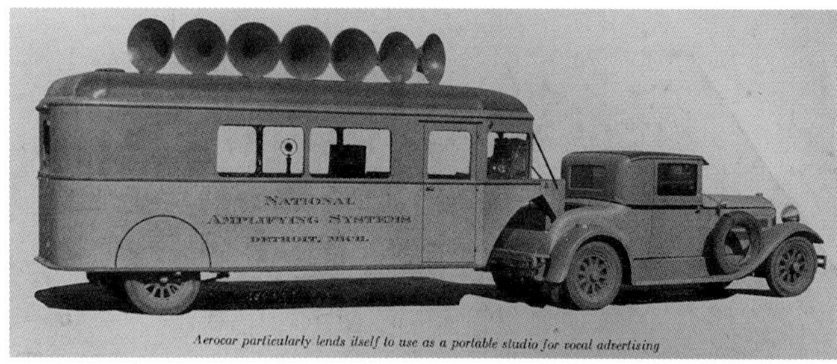

National Amplifying Systems (1930s)

Fessenden White (1930s)

San Antonio Drug Company (1930s)

Old Ben Coal Corporation (1930s)

Pan American Airways (1930)

Clark's Top and Body Works (1930)

Enna Jettick (1930)

Friendly Five Shoes (1930)

Lail Brothers (1930)

Philbrick Ambulance (1930)

Standard Talking Machine (1930)

Carl Adams Builder (1931)

John Housley Cigar Company (1931)

Norge Electric Refrigerators (1931)

General Electric Company (1932)

Florida Year Round Clubs (1933)

Roosevelt Warm Springs institute (1933)

Fostoria Glass Co. (1934)

Cities Services Company (1934)

Crosley Radios (1935)

Activated Alum Corp (1936)

Brooklyn Borough Gas Company (1937)

Grove Laboratories (1937)

Modern Equipment Corporation (1937)

Servel Electrolux (1937)

Wayside Cathedral (1939)

APPENDIX F

A BRIEF HISTORY OF THE FIFTH WHEEL

The Auto Salon de Luxe fifth-wheel recreational vehicle built for Baron Crawhez by Auto-Mixte Pescatore of Belgium (1912). *Courtesy of Michel Bedeur*

The significance of the Aerocar is due in part to its design as a fifth-wheel trailer. A short overview of the history, purpose, and benefits of the fifth wheel may assist in gaining a better appreciation of this revolutionary vehicle.

A device called a "wheel plate" was developed in Europe in about the eighteenth century for use in horse-drawn carriages. It was placed in the middle of the carriage's front axle and had two circular iron plates with central holes that received a bolt or pin. It allowed an axle to pivot horizontally at the front of a chassis for steering purposes. Wheel plates were large, round, and heavy and often covered in horse fat to reduce friction between the plates.

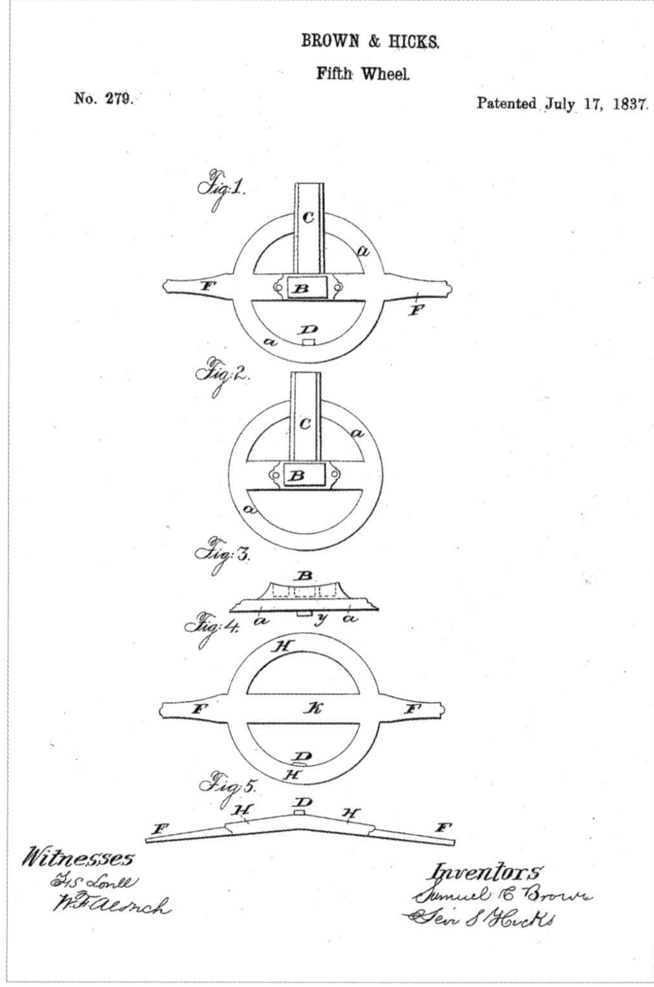

The Brown & Hicks fifth-wheel patent (1837)

The device known as the fifth wheel was developed in the US as a lightweight replacement for the wheel plate. American horse-drawn carriages needed to be lighter for travel over rougher ground than in Europe. It was probably invented in the early 1800s, but nobody knows who first invented them, since most US patents issued between 1790 and 1836 were destroyed by a patent office fire in 1836.

Every known fifth-wheel patent issued from 1836 onward (there are hundreds) has been a lightweight refinement of the wheel plate. Improvements of the basic design over time helped reduce friction, improve turning circles, and stabilize the carriage on uneven surfaces by incorporating flexing elements or springs. Over time, fifth wheels were strengthened as loads increased and carriage construction methods evolved.

FIFTH-WHEEL GOODS TRAILERS

With the advent of motorized road transportation in the US at the start of the twentieth century, horses were gradually replaced by horsepower, and fifth wheels took on a new purpose in the goods-trucking industry. They were used to join a "tractor" (a motorized tow vehicle) to a single-axle trailer that came to be known as a semitrailer. The front axle of a standard twin-axle trailer was removed altogether, and its front end was placed on the tractor. To allow the semitrailer to pivot, a modified fifth wheel was used.

Because the semitrailer had lost its front axle, the tractor needed to assume about one-third of the trailer's weight. To do so without causing the tractor's front end to lift required a fifth wheel that was located far enough forward on the tractor, typically over or slightly ahead of the tractor's rear axle. Both the fifth wheel and the tractor's real axle had to be strengthened to assume new load-bearing responsibilities.

In the early 1900s, many new patents were issued for semitrailer fifth wheels drawn by motorized tractors. As semitrailer fifth wheels evolved, they incorporated a range of self-locking, quick-release, and shock-absorbing mechanisms.

FIFTH-WHEEL ADVANTAGES

The advantages of a fifth-wheel trailer over a conventional trailer were initially seen as a tighter turning circle and the ability to swap over trailer or tractor quickly. These were important benefits in the goods transport industry. As vehicle speed increased, it became apparent that fifth-wheelers had another key benefit over the more traditional trailer hitch. Because the fifth-wheel connection point was located above rather than behind the tractor's rear axle, a fifth-wheel trailer was far more stable at speed. Trailer sway (or, more technically, yaw) was reduced, and greater weights could be towed at higher speeds compared to conventionally hitched trailers. In the early twentieth century, fifth wheels were also called "turntables" in the goods industry, a label that emphasized the tight turning circles achievable by these devices.

THE CURTISS CONTRIBUTION TO FIFTH WHEELS

Glenn Curtiss did not invent the fifth wheel, but he was the first to use a fifth-wheel trailer for recreational purposes in America. The first recreational fifth-wheeler in the world came from Belgium in 1912. Curtiss first used a fifth wheel on the Curtiss Camp Car in about 1919. It is reported that he got the idea of using a fifth-wheel trailer for recreational purposes by observing how well aircraft rode when being towed backward at airfields, with their rear ends sitting on top of an automobile or truck.

The Aero Coupler fifth wheel as used on the Aerocar from 1928 onward was a Curtiss invention. It was a further refinement of existing fifth wheels, incorporating for the first time a pneumatic shock absorber into the fifth-wheel receiver of the tow vehicle in the form of a horizontally placed aircraft wheel.

Curtiss was one of the first road vehicle designers to understand the stability benefits of fifth wheels some twenty-five years before towing dynamics were formally studied and understood in the 1950s. The high speeds at which the Aerocar could be towed, often up to 70 mph (113 km/h), meant that a good understanding of towing dynamics was essential to avoid accidents. As such, Curtiss was a pioneer in trailer safety.

SOURCES AND BIBLIOGRAPHY

Some of the busiest people working in Florida during the 1920s and '30s were the secretaries (no doubt there were more than one) of Miami Beach founder Carl G. Fisher. Fisher dictated thousands of pages of letters sent to his friends and business associates across America as he promoted his ideas and products. We have Fisher to thank for documenting the early days of the Aerocar story, and a number of institutions to thank for preserving his letters for us to read today. There are four places that I know of that have Fisher correspondence: the Carl Graham Fisher Papers at HistoryMiami Museum, the Aerocar Corp. files within the Clement Melville Keys Papers in the National Air and Space Museum Archives, the Glenn H. Curtiss Collection in the National Air and Space Museum Archives, and the Glenn H. Curtiss Museum archives. The first three of these have been digitized. All have been valuable resources for this book.

Clearly the Curtiss and Keys Papers at the National Air and Space Museum are important resources in their own right, as is just about everything in the Curtiss Museum archives. The letters written by Curtiss himself are historically significant as prime source material. I look forward to the day when the Curtiss Museum has the financial and staffing resources to digitize its important Curtiss archives.

Curtiss Museum curator emeritus Rick Leisenring Jr. wisely suggested that I tread with care when reading Curtiss biographies. Much fiction surrounding Curtiss's life has become fact, and quotes by others have become quotes by Curtiss. An obvious example is *The Curtiss Aviation Book* (New York: Stokes, 1912), written by "Glenn H. Curtiss and Augustus Post" but in fact written almost entirely by Post and others. The most valuable Curtiss "biography" I found to be Cecil R. Roseberry's *Glenn Curtiss: Pioneer of Flight* (Garden City, NY: Doubleday, 1972). Although like most other biographies it covers mostly Curtiss's flying achievements, it is both detailed and reasonably accurate.

The digitization of early newspapers has continued apace, and I found www.newspapers.com to be a particularly useful source for Florida-based news. They are to be commended for their extensive collection. The best short summary of the Aerocar is, in my view, Roger B. White's "Planes, Trailers and Automobiles: The Land Yachts of Glenn Curtiss," *Automobile Quarterly* 32, no. 3 (April 1994), which can be found online. Mark Theobald also has an excellent overview of the company's history at www.coachbuilt.com.

Other consulted material:

Ballinger, Kenneth. *Miami Millions: The Dance of the Dollars in the Great Florida Land Boom of 1925*. Miami, FL: Franklin, 1936.

Bramson, Seth H. *The Curtiss-Bright Cities: Hialeah, Miami Springs and Opa Locka*. Charleston, SC: History Press, 2008.

Brimmer, F. E. *Autocamping*. Cincinnati: Stewart Kidd, 1923.

Casey, Louis S. *Curtiss: The Hammondsport Era, 1907–1915*. New York: Crown, 2003.

Curtiss, Glenn H., and Augustus Post. *The Curtiss Aviation Book*. New York: Frederick A. Stokes, 1912.

George, Paul S. "Passage to the New Eden: Tourism in Miami from Flagler through Everest G. Sewell." *Florida Historical Quarterly* 59, no. 4 (1980): 440–63.

Hatton, Hap. *Tropical Splendor: An Architectural History of Florida*. New York: Alfred A. Knopf, 1987.

House, Kirk W. *Hell-Rider to King of the Air*. Warrendale, PA: SAE International, 2003.

Long, John Cuthbert, and John Dietrich Long. *Motor Camping*. New York: Dodd, Mead, 1923.

Nelson, Richard Allen. "Palm Trees, Public Relations, and Promoters: Boosting Florida as a Motion Picture Empire, 1910–1930." *Florida Historical Quarterly* 61, no. 4 (1983): 383–403.

Prentice, Colonel James. "The Influence of Aircraft Design on the Trend of Motor Vehicle Construction." *U.S. Air Services*, December 1930.

Roseberry, Cecil R. *Glenn Curtiss: Pioneer of Flight*. Garden City, NY: Doubleday, 1972.

Scharff, Robert, and Walter S. Taylor. *Over Land and Sea: A Biography of Glenn Hammond Curtiss*. New York: D. McKay, 1968.

Studer, Clara. *Sky Storming Yankee: The Life of Glenn Curtiss*. New York: Stackpole Sons, 1937.

Theobald, Mark. "Curtiss Aerocar Co.—Aerocar Co. of Detroit—Adams Trailer Corp." www.coachbuilt.com/bui/c/curtiss.

Trimble, William F. *Hero of the Air: Glenn Curtiss and the Birth of Naval Aviation*. Annapolis, MD: Naval Institute Press, 2010.

Vergara, George L. *Hugh Robinson, Pioneer Aviator*. Gainesville: University Press of Florida, 1995.

White, Roger B. *Home on the Road: The Motor Home in America*. Washington, DC: Smithsonian Institution Press, 2000.

White, Roger B. "Planes, Trailers and Automobiles: The Land Yachts of Glenn Curtiss." *Automobile Quarterly* 32, no. 3 (April 1994).

Zobel, Myron. *The 14-Karat Trailer*. New York: Frederick Fell, 1955.

INDEX

ABOUT THE AUTHOR

Andrew Woodmansey is a recreational-vehicle historian based in Sydney, Australia. He has been RV-ing since he was a child. His first RV history book, *Recreational Vehicles: A World History, 1872–1939*, was published in January 2022. He also blogs at rvhistory.com. Prior to specializing in RV history, he self-published *Caravan Buyers Guide* for the Australian market. This gave him a detailed knowledge of RV technology and towing dynamics that is important for an understanding and explanation of the technical features incorporated into the Curtiss Aerocar. He currently owns a travel trailer that he uses to explore Australia.

Curtiss